Transportation Policy 2018

AA005506

Transportation Research Record: Journal of the Transportation Research Board Volume 2672, Issue 7

ISBN: 978-1-5108-8177-8

Printed from e-media with permission by:

Curran Associates, Inc.
57 Morehouse Lane
Red Hook, NY 12571

Some format issues inherent in the e-media version may also appear in this print version.

Published by Sage Publications USA

Copyright© (2018) by Transportation Research Board of the National Academies
All rights reserved.

ISBN (Print) 978-1-5108-8177-8
ISBN 2018 Printed Set (All Issues) 978-1-5108-7735-1

Printed by Curran Associates, Inc. (2019)

For permission requests, please contact sagepub.com/journals-permissions

Additional copies of this publication are available from:

Curran Associates, Inc.
57 Morehouse Lane
Red Hook, NY 12571 USA
Phone: 845-758-0400
Fax: 845-758-2633
Email: curran@proceedings.com
Web: www.proceedings.com

Volume 2672 Number 7 December 2018

TRR
JOURNAL OF THE TRANSPORTATION RESEARCH BOARD

Contents

Transportation Policy

Article

2018 CHARLEY V. WOOTAN AWARD: Outstanding Paper in Policy and Organization

Articles

TRANSPORTATION RESEARCH BOARD 2019 EXECUTIVE COMMITTEE*

OFFICERS

CHAIR: **Victoria A. Arroyo**, *Executive Director, Georgetown Climate Center; Assistant Dean, Centers and Institutes; and Professor and Director, Environmental Law Program, Georgetown University Law Center, Washington, D.C.*
VICE CHAIR: **Leslie S. Richards**, *Secretary, Pennsylvania Department of Transportation, Harrisburg*
EXECUTIVE DIRECTOR: **Neil J. Pedersen**, *Transportation Research Board*

MEMBERS

Michael F. Ableson, *Vice President, Global Strategy, General Motors Company, Detroit, MI*
Carlos M. Braceras, *Executive Director, Utah Department of Transportation, Salt Lake City*
Ginger Evans, *President, Tower Consulting, LLC, Arlington, VA*
Nuria I. Fernandez, *General Manager/CEO, Santa Clara Valley Transportation Authority, San Jose, CA*
Nathaniel P. Ford Sr., *Executive Director–CEO, Jacksonville Transportation Authority, Jacksonville, FL*
A. Stewart Fotheringham, *Professor, School of Geographical Sciences and Urban Planning, Arizona State University, Tempe*
Susan Hanson, *Distinguished University Professor Emerita, Graduate School of Geography, Clark University, Worcester, MA*
Stephen W. Hargarten, *Professor, Emergency Medicine, Medical College of Wisconsin, Milwaukee*
Chris T. Hendrickson, *Hamerschlag University Professor of Engineering, Carnegie Mellon University, Pittsburgh, PA*
S. Jack Hu, *Vice President for Research and J. Reid and Polly Anderson Professor of Manufacturing, University of Michigan, Ann Arbor*
Roger B. Huff, *President, HGLC, LLC, Farmington Hills, MI*
Ashby Johnson, *Executive Director, Capital Area Metropolitan Planning Organization (CAMPO), Austin, TX*
Geraldine Knatz, *Professor, Sol Price School of Public Policy, Viterbi School of Engineering, University of Southern California, Los Angeles*
William Kruger, *Vice President, UPS Freight for Fleet Maintenance and Engineering, Richmond, VA*
Michael R. McClellan, *Vice President, Strategic and Network Planning, Norfolk Southern Corporation, Norfolk, VA*
Patrick K. McKenna, *Director, Missouri Department of Transportation, Jefferson City*
Brian Ness, *Director, Idaho Transportation Department, Boise*
Susan A. Shaheen, *Adjunct Professor, Co-Director, Transportation Sustainability Research Center, University of California, Berkeley*

James M. Tien, *Distinguished Professor and Dean Emeritus, College of Engineering, University of Miami, Coral Gables, FL*
Shawn Wilson, *Secretary, Louisiana Department of Transportation and Development, Baton Rouge*

EX OFFICIO MEMBERS

Ronald Batory, *Administrator, Federal Railroad Administration, U.S. Department of Transportation*
Michael R. Berube, *Acting Assistant Secretary for Sustainable Transportation, U.S. Department of Energy, Washington, D.C.*
Mark H. Buzby *(Rear Admiral, U.S. Navy), Administrator, Maritime Administration, U.S. Department of Transportation*
Steven Cliff, *Deputy Executive Officer, California Air Resources Board, Sacramento*
Edward N. Comstock, *Independent Naval Architect, Sunbury, MD*
Howard R. Elliott, *Administrator, Pipeline and Hazardous Materials Safety Administration, U.S. Department of Transportation*
Daniel K. Elwell, *Acting Administrator, Federal Aviation Administration, U.S. Department of Transportation*
LeRoy Gishi, *Chief, Division of Transportation, Bureau of Indian Affairs, U.S. Department of the Interior, Germantown, MD*
John T. Gray II, *Senior Vice President, Policy and Economics, Association of American Railroads, Washington, D.C.*
Brandye Hendrickson, *Deputy Administrator, Federal Highway Administration, U.S. Department of Transportation*
Nikola Ivanov, *Director of Operations, Center for Advanced Transportation Technology Laboratory, University of Maryland, College Park, and Chair, TRB Young Members Council*
Heidi King, *Deputy Administrator and Acting Administrator, National Highway Traffic Safety Administration, U.S. Department of Transportation, Washington, D.C.*
Raymond Martinez, *Administrator, Federal Motor Carrier Safety Administration, Washington, D.C.*
Keith Nelson, *Senior Advisor to the Secretary, Office of the Assistant Secretary for Research and Technology, U.S. Department of Transportation*
Craig A. Rutland, *U.S. Air Force Pavement Engineer, U.S. Air Force Civil Engineer Center, Tyndall Air Force Base, FL*
Karl Simon, *Director, Transportation and Climate Division, U.S. Environmental Protection Agency*
Paul Skoutelas, *President and CEO, American Public Transportation Association, Washington, D.C.*
Scott A. Spellmon *(Major General, U.S. Army), Deputy Commanding General for Civil and Emergency Operations, U.S. Army Corps of Engineers*
Katherine F. Turnbull, *Executive Associate Director and Research Scientist, Texas A&M Transportation Institute, College Station*
Jim Tymon, *Executive Director, American Association of State Highway and*

Transportation Officials, Washington, D.C.
K. Jane Williams, *Acting Administrator, Federal Transit Administration, U.S. Department of Transportation*
Paul Zukunft *(Admiral, U.S. Coast Guard), Commandant, U.S. Coast Guard, Washington, D.C.*

* Membership as of February 2019.

TRB

TRANSPORTATION RESEARCH BOARD

CORE SPONSORS

Transportation Departments of the 50 States and the District of Columbia

Federal Government

U.S. Department of Transportation
Federal Aviation Administration (FAA)
Federal Highway Administration
Federal Motor Carrier Safety Administration (FMCSA)
Federal Railroad Administration (FRA)
Federal Transit Administration
National Highway Traffic Safety Administration (NHTSA)
Office of the Assistant Secretary of Transportation for Research and Technology (OST-R)

U.S. Air Force Civil Engineer Center (AFCEC)
U.S. Coast Guard (USCG)
U.S. Department of Energy
U.S. Department of the Interior (DOI)
U.S. Environmental Protection Agency (EPA)

Nongovernmental Organizations

American Public Transportation Association (APTA)
Association of American Railroads (AAR)
California Air Resources Board

The National Academies of
SCIENCES • ENGINEERING • MEDICINE

The **National Academy of Sciences** was established in 1863 by an Act of Congress, signed by President Lincoln, as a private, nongovernmental institution to advise the nation on issues related to science and technology. Members are elected by their peers for outstanding contributions to research. Dr. Marcia McNutt is president.

The **National Academy of Engineering** was established in 1964 under the charter of the National Academy of Sciences to bring the practices of engineering to advising the nation. Members are elected by their peers for extraordinary contributions to engineering. Dr. C. D. Mote, Jr., is president.

The **National Academy of Medicine** (formerly the Institute of Medicine) was established in 1970 under the charter of the National Academy of Sciences to advise the nation on medical and health issues. Members are elected by their peers for distinguished contributions to medicine and health. Dr. Victor J. Dzau is president.

The three Academies work together as the National Academies of Sciences, Engineering, and Medicine to provide independent, objective analysis and advice to the nation and conduct other activities to solve complex problems and inform public policy decisions. The Academies also encourage education and research, recognize outstanding contributions to knowledge, and increase public understanding in matters of science, engineering, and medicine.

Learn more about the National Academies of Sciences, Engineering, and Medicine at **www.national-academies.org.**

The **Transportation Research Board** is one of seven major programs of the National Academies of Sciences, Engineering, and Medicine. The mission of the Transportation Research Board is to increase the benefits that transportation contributes to society by providing leadership in transportation innovation and progress through research and information exchange, conducted within a setting that is objective, interdisciplinary, and multimodal. The Board's varied activities annually engage about 7,000 engineers, scientists, and other transportation researchers and practitioners from the public and private sectors and academia, all of whom contribute their expertise in the public interest. The program is supported by state transportation departments, federal agencies including the component administrations of the U.S. Department of Transportation, and other organizations and individuals interested in the development of transportation.

Learn more about the Transportation Research Board at **www.TRB.org**.

Article

TRR
JOURNAL OF THE TRANSPORTATION RESEARCH BOARD

Automated Vehicle Legislative Issues

Transportation Research Record
2018, Vol. 2672(7) 1–13
© National Academy of Sciences:
Transportation Research Board 2018
Article reuse guidelines:
sagepub.com/journals-permissions
DOI: 10.1177/0361198118774155
journals.sagepub.com/home/trr
SAGE

Sarah M. L. Hubbard[1]

Abstract

This paper provides a synthesis of issues addressed by state legislation in relation to automated vehicles (AV); technologies are rapidly evolving and many states have developed legislation to govern AV testing and deployment and assure safety on public roads. The expected benefits of AV include increased safety, increased capacity, decreased congestion, increased productivity because of hands-free travel, and increased mobility for people who cannot drive. The projected economic impact of AV is significant, with an estimated market of $7 trillion by 2050. Potential challenges include increased costs, liability and licensing issues, privacy considerations, security and cybersecurity concerns, and potential job losses. Legislative responses to AV technology have varied significantly, depending on the state. Twenty-one states and the District of Columbia have passed legislation related to AV, and six states have executive orders. Even after federal AV legislation is passed, there will still be an important role for states in AV licensure, registration, insurance, traffic laws, enforcement, infrastructure, and emergency response. The objective of this research is to identify and synthesize current state legislation related to AV, which shows diverse responses and has led to a patchwork of operating conditions from state to state. The information presented in this paper provides important information as agencies and decision makers develop strategic plans for AV activities at every level, and serves an important function by documenting the evolution of issues related to AV which is an important part of transportation history.

Numerous states have enacted legislation to assure a safe and well-defined framework for automated vehicle (AV) technologies. AV technologies are rapidly evolving and are expected to provide many benefits including increased safety, increased capacity, decreased congestion, increased productivity because of hands-free travel, and increased mobility for people unable to drive. The projected economic impact of AV is $1.2 trillion or $3,800/person per year (*1*), with an estimated market of $7 trillion by 2050 (*2*). Although speculative, this indicates the dramatic impact of AV on transportation. The benefits may be significant; however, there are also potential challenges, including increased costs, liability and licensing issues, security and privacy concerns (*3*), cybersecurity threats, and potential job losses.

State AV legislation has varied significantly. Currently 21 states and the District of Columbia have passed AV legislation, and six states have executive orders. The objective of this research is to identify and synthesize current state legislation related to AV. After providing background information on AV, including definitions for levels of automation, this document identifies states with AV legislation and executive orders, and presents a discussion of legislative trends. Subsequent sections discuss how state legislation addresses nine legislative topic areas (e.g., licensing and registration, operating requirements and infrastructure) and other considerations, including potential implementation problems associated with differences in state laws. The final section, Conclusions, highlights the main discussion points and provides a context for future activities.

Background

The legislative framework for AV is still evolving at the national and state level. At the national level, the National Highway Traffic Safety Administration (NHTSA) published the *Federal Automated Vehicles Policy* (FAVP) in 2016 (*4*). This document provides guidance rather than regulations, and sets forth a standard framework for levels of autonomy, vehicle performance, a model state policy, and current and future regulatory tools. In September 2017, NHTSA published *A Vision for Safety 2.0*, which reinforces the voluntary nature of previous guidelines, supports the safe deployment of automated driving systems (ADS), and clarifies state and federal roles (*5*).

The levels of autonomy shown in Table 1 provide common terminology for agencies and industry. Level 0 (L0) reflects no automation and level 5 (L5) reflects full automation. As automation increases, the responsibility for operation transitions from the human operator to the AV. The

[1]Purdue University, West Lafayette, IN

Corresponding Author:
Address correspondence to Sarah M. L. Hubbard: sarahh@purdue.edu

Table 1. SAE International Definitions for Levels of Autonomy

Responsible for monitoring environment	Level	Description
Human operator	0	No automation, human driver does everything
	1	Driver assistance, ADAS provides limited assistance with some functions (e.g., backup warning beep)
	2	Partial automation, ADAS can conduct some parts of driving task, human monitors and controls other tasks (e.g., cruise control)
Automation (HAV = highly automated vehicle)	3	Conditional automation, vehicle ADS can conduct some parts of task and monitor driving environment but human ready to take back control
	4	High automation, vehicle ADS can conduct driving task and monitor driving environment but only in certain environments and under certain conditions
	5	Full automation, vehicle ADS performs all tasks under all conditions

Source: (*4*).

primary distinction between conventional vehicle operation (L0, L1, and L2) and a highly automated vehicle (HAV; L3, L4, and L5) is the responsibility for monitoring the environment. This document uses the term HAV for L3, L4, and L5 per the FAVP convention; HAV is not an acronym for "high automation," which is the name for level 4 as referenced by SAE. These categories are consistent with *A Vision for Safety 2.0* (*5*), which uses the terminology advanced driver assistance system (ADAS) for L1 and L2, and ADS for L3, L4, and L5.

It is likely that L4, automated operation on a freeway during good weather, may be realized before L5, automated operation on any public street in any weather condition. Some companies plan to skip L3 altogether, because of concerns with driver vigilance and potential challenges with human drivers being able to respond quickly and appropriately when needed (*6*).

Level 5 includes operation in urban areas with a variety of users, including pedestrians, cyclists, and delivery trucks. Level 5 also includes operation through complex geometries, such as one-way street networks and roundabouts, and can adapt to local regulations that may vary depending on jurisdiction and time of day, including local speed limits and school zones, and peak-hour turn prohibitions.

The second part of the FAVP is guidance for vehicle performance, intended to assure safety during testing and provide data that may be useful for the development of future regulations. Vehicle safety standards are currently the responsibility of the federal government, and all vehicles sold in the U.S. must meet Federal Motor Vehicle Safety Standards (FMVSS). The FAVP proposes that the federal government maintain this responsibility, and outlines a voluntary 15-point safety assessment. The safety assessment addresses safety considerations (e.g., crashworthiness and driver training), legal considerations (registration and certification; federal, state, and local laws), considerations related to data (data recording, sharing, privacy and cybersecurity), and automated functions (operational design domain, object detection and response, minimal risk condition, and validation methods). Manufacturers must obtain an exemption to FMVSS if vehicles do not meet current standards—for example, if they do not have a steering wheel, accelerator, or brake pedal. Currently, NHTSA grants exemptions for up to 2,500 vehicles per year. Federal legislation has been proposed to increase this limit to facilitate more rapid innovation.

The third component of the FAVP is the model state policy for states that regulate AV testing and operation; this clarifies federal and state AV responsibilities and supports seamless operation between states. In addition to safety standards, the federal government is responsible for compliance, vehicle recalls, public information about safety issues, and guidance to support national safety goals. According to the FAVP, state AV responsibilities include:

- Driver licensing and vehicle registration,
- Traffic laws and enforcement,
- Safety inspections, and
- Motor vehicle insurance and liability.

Best practices for legislatures are also addressed by *A Vision for Safety 2.0*.

In September 2017, the US House passed the Safely Ensuring Lives Future Deployment and Research in Vehicle Evolution Act (SELF DRIVE) (*7*) which:

- Expands federal preemption beyond safety to include HAV operation (this expansion contradicts the traditional framework in which states license vehicle drivers),
- Updates the FMVSS to reflect the changes required because of the inclusion of HAV,
- Increases FMVSS exemptions from the current limit of 2,500 per year up to 100,000 per year per manufacturer (with an incremental increase over time), and
- Establishes an advisory council on HAV.

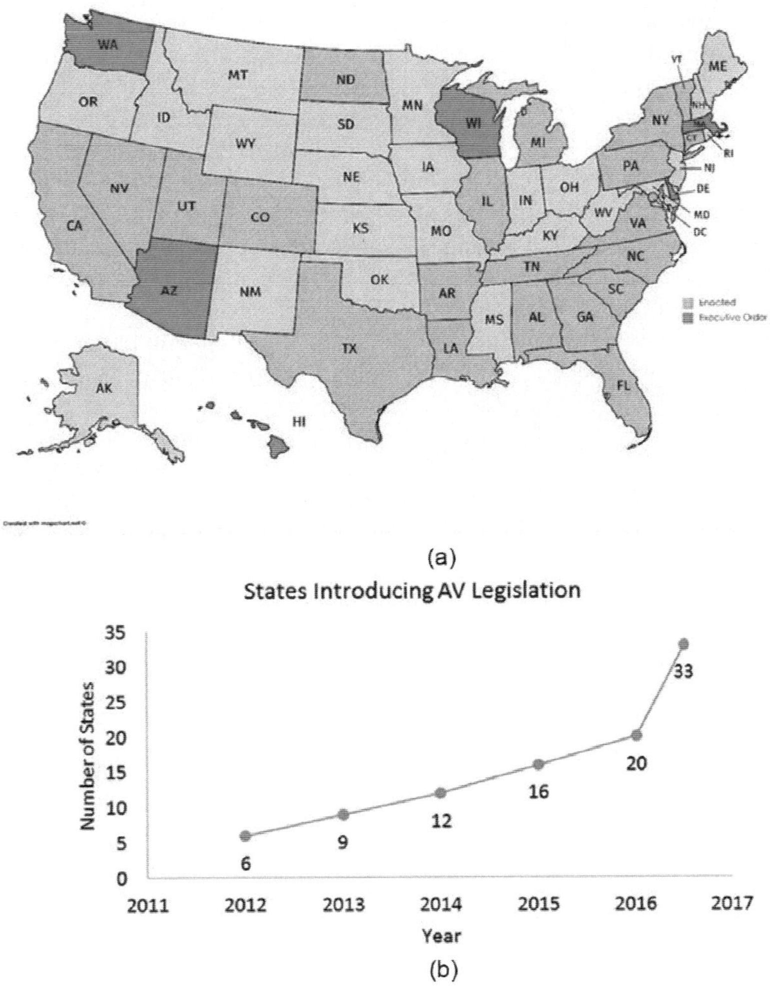

Figure 1. State AV legislation, showing (*a*) states with AV legislation (as of December 1, 2017), and (*b*) increase in state AV legislation introduced (as of December 1, 2017).
Source (*10*).

In October 2017, the US Senate Commerce, Science and Transportation Committee passed the American Vision for Safer Transportation through Advancement of Revolutionary Technologies Act (AV START) (*8*). The AV START Act is similar to SELF DRIVE Act in many respects. Both maintain common law liability under state law, limit state regulations in some areas related to HAV, exclude large commercial vehicles from the legislation, and increase the FMVSS exemptions to allow more vehicle sales, with incremental increases over a 3-year period. AV START defers safety review and approval through the federal motor carrier process for 3 years, requires manufacturers to submit a safety evaluation report to the Department of Transportation (DOT), and focuses more attention on data use, privacy, and cybersecurity, as evidenced by advisory committees addressing technical issues and data access, and a work group focused on consumer education.

Automated Vehicle State Legislation

Twenty-one states and the District of Columbia have enacted legislation and six governors have issued executive orders, as shown in Figure 1*a*. Nevada was the first state to enact legislation in 2011; this legislation provided definitions, authorized operation, and directed the Department of Motor Vehicles (DMV) to develop rules for licensure and operation (*9*). Subsequent legislation has varied significantly. In some states, legislation is minimal and may focus on definitions. In other states, test programs are authorized and operational limits are specified.

States that were early adopters have often passed new legislation to modify the original framework for AV, reflecting the rapid technology advances and a changing context for AV operation. The need for flexibility can also be addressed with sunset provisions and delegation of responsibilities to state agencies, which allows programs to be modified, as warranted.

Legislative activity is ramping up, and 22 AV bills were enacted in 15 states in 2017 (*10*). According to the National Conference of State Legislature (NCSL), there were 101 proposed bills in 33 states related to AV in 2017 (*10*). Most of these (97) were proposed in the first half of the year, reflecting the way many state legislatures begin their session in January and have a limited number of legislative days. This is a dramatic increase from previous years, as shown in Figure 1*b*.

Legislation and its impact varies significantly. State legislation and executive orders were reviewed to determine whether they addressed the following topic areas:

- Definitions,
- Study request,
- Licensing and registration,
- Insurance and liability,
- Operator requirements,
- Infrastructure,
- Vehicle testing and operation,
- Commercial vehicle operation (CVO),
- Privacy.

The results are shown in Table 2 and discussed in the following sections. Discussion of these topic areas, which were developed based on the 12 topic areas defined by the National Conference of State Legislatures (*10*), is intended to provide an overview of how topics are addressed with examples, rather than represent a complete inventory of every legislative component in each state. In some cases, there is overlap between topic areas: for example, often operator requirements may overlap with the requirements for license, registration, and insurance. The assessment in this document does not align directly with the search results by topic area obtained using the NCSL website because the NCSL provides categorization at a higher level and this assessment considers more detailed components of the legislation.

Definitions

Definitions are one of the most basic components of legislation and may stand alone (e.g., Louisiana [*11*]), or be part of more comprehensive legislation. Common terms and concepts defined include autonomous vehicle, autonomous driving system, autonomous technology, operational design domain, minimal risk condition, operator or human operator, and platoon.

Legislation has defined AV terms differently as activities evolve and technologies progress. The publication of the FAVP has led to increased consistency for definitions, and many states have included references to the SAE levels of automation. Nevada specifies that AV means L3, L4, or L5 (*12*). Colorado specifies that L0 to L3 is not addressed by the legislation, and defines an ADS as HAV L4 or L5 (*13*). Connecticut defines SAE J3016 but does not refer to levels of automation or how these levels translate to requirements in the state (*14*). The same concept may be referred to by different names in different states. A platoon may also be called a coordinated platoon (Georgia [*15*]), a driver-assistive truck platooning system (Arkansas [*16*] and Florida [*17*]) or a connected braking system (Texas [*18*]).

Definitions may also clarify what is excluded. Numerous states (e.g., Florida [*19*], California [*20*], and Tennessee [*21*]) clarify that a vehicle is not an AV because it has systems that enhance safety or provide driver assistance, as long as the technologies are not capable of operating the vehicle or monitoring by the human operator. Excluded technologies typically include collision avoidance, electronic blind spot assistance, adaptive cruise control, and other L1 and L2 technologies.

Texas defines the entire dynamic driving task in terms of operational aspects (steering, braking, and monitoring environment) and tactical aspects (responding to events and determining when to take action), and clarifies that it does not include strategic aspects (determining destinations) (*22*); analogous definitions are also seen in Louisiana (*11*) and Connecticut (*14*).

As technologies progress, legislation looks forward to future applications and defines new terms that may apply when there is no human driver, and when taxi or ridesharing fleets consist of AV. Michigan defines the SAVE program (*23*), which is similar to the autonomous vehicle network company defined in Nevada (*12*). Definitions are important and may provide clarification, but, in some cases, the lack of a consistent, standard and shared vocabulary can present obstacles, and legislative attempts to define even basic terms such as operator and driver can be confusing (*24*).

Study Request

Study requests and committee appointments are a common component of AV legislation. Five executive orders appoint an AV work group, for example (*25–29*). In Alabama (*30*), North Dakota (*31*, *32*), Utah (*33*, *34*) and Vermont (*35*), this is the entirety of the legislation. Alabama established a joint legislative committee to study all aspects including safety and economic impacts, and report back in a year, at which point the committee would be dissolved. North Dakota's legislation requested a study of AV issues such as legislation, licensing, registration, insurance, and inspection, as well as data ownership and potential uses for the information gathered by AV.

In other states, studies, committees and reports are one component of broader legislation. Florida's 2012 legislation provided a framework for AV issues and also required that state agencies submit a report about the need for additional regulations or legislation for safe testing and operation (*19*). Legislation in both Florida (*17*) and Utah (*34*) required a

Table 2. Topics Addressed by State AV Legislation

State	Year	Definitions	Study	License and Registration	Insurance and Liability	Operator Requirements	Infrastructure	Vehicle Testing and Operation	CVO	Privacy
Enacted legislation										
AL	2016		✓							
AR	2017	✓						✓	✓	
CA	2012, 2016	✓	✓	✓	✓	✓		✓		✓
CO	2017	✓	✓					✓		
CT	2017	✓	✓	✓	✓	✓		✓		
FL	2012 2016	✓	✓	✓	✓	✓	✓	✓	✓	
GA	2017	✓		✓	✓	✓		✓	✓	
IL	2017	✓						✓		
LA	2016	✓								
MI	2013, 2016	✓	✓	✓	✓	✓		✓	✓	✓
NV	2011, 2013, 2017	✓		✓	✓	✓			✓	✓
NY	2017	✓	✓	✓	✓	✓		✓		
NC	2017	✓	✓	✓	✓	✓		✓	✓	
ND	2015, 2017		✓							
PA	2016						✓			
TN	2015, 2016, 2017	✓	✓	✓	✓	✓		✓	✓	
TX	2017	✓		✓	✓	✓		✓	✓	
UT	2015, 2016	✓	✓					✓	✓	
VA	2016			✓				✓		
VT	2017		✓							
DC	2012	✓			✓	✓				
Executive orders										
AZ	2015		✓	✓	✓	✓		✓		
DE	2017		✓							
HI	2017							✓		
MA	2016		✓		✓	✓		✓		
WA	2017		✓	✓	✓	✓		✓		
WI	2017		✓							

study on truck platooning. Colorado's 2017 legislation mandates that the DOT provide an annual report about AV testing in Colorado (*13*).

Study requests often reflect future policy. Michigan's 2013 legislation required recommendations for legislation relating to liability (*36*), which was later addressed by SB 663 (2016) (*37*), which limited liability for original manufacturers in the case of conversion by upfitters. In 2016, Michigan also legislated the Michigan council on future mobility within the DOT to provide recommendations for policy changes on an annual basis (*23*).

Licensing and Registration

Legislation

Vehicle licensing and registration is a well-defined process for traditional motor vehicles in all states, and assures that states have information about vehicles and owners and that vehicles are insured, as well as providing a source of revenue.

Licensing and registration for AV is addressed by legislation in numerous states. In some cases, special plates are required for AV (e.g., Michigan [*37*]); in other states, such as Virginia (*38*), AV that are being tested are issued standard plates to assure that other drivers do not change their behavior and affect testing results (*39*). In Nevada, legislation specifies that the Department regulations include provisions related to license plates and registration but these should not unusually impede AV testing and operation (*12*). Some legislation states that an owner may identify the vehicle as automated (e.g., Texas [*22*]), whereas other legislation requires registration as automated (e.g., Georgia [*40*], Tennessee [*41*]).

States may reiterate the need for vehicles to be compliant with applicable federal standards and regulations for registration (e.g., Florida [*42*]), with recent legislation more likely to require compliance with standards established by the NHTSA for fully autonomous vehicles. These requirements may be included as part of the legislation related to registration, as well as the legislation related to operators, testing, and operational requirements, as discussed in later sections.

Fees and Funding

Fees for vehicle licensing and registration are often an important source of revenue for roadway infrastructure. The location data associated with AV may facilitate new funding opportunities, including user charges on a per mile basis, which could be adjusted to reflect demand and reduce congestion. No legislation has included this kind of fee, although early versions of legislation in Tennessee did include a mileage-based fee for AV, which was removed in the version that passed (*21*). Nevada specifies an excise tax for autonomous vehicle network companies (*12*), with a 3% tax on the total fare. A fee to cover administrative costs for authorization of an AV is often authorized as well (e.g., California [*20*]). Some states specify that local jurisdictions cannot charge fees (e.g., Michigan [*23*] and Nevada [*12*]), although Nevada allows a local jurisdiction, including airports, to charge a fee for a business license if this is consistent with standard charges for permits.

Insurance and Liability

Insurance and liability is addressed by all states for traditional motor vehicles, and is also a common topic for AV legislation. The first AV legislation in Nevada directed the DMV to adopt rules for insurance and safety standards (*9*). Subsequent Nevada legislation required proof of $5 million liability insurance for testing (*43*), which has been maintained in the most recent legislation (*12*) and is a common requirement in other states, as well (e.g., California [*44*], Connecticut [*14*] New York [*45*]).

Texas requires that AVs carry the same insurance coverage as other vehicles (*22*), and Colorado (*13*) and Tennessee (*41*) specify that the liability is consistent with state law, federal law, or common law. Insurance requirements may vary depending on the level of automation or operating framework. Georgia specifies that AV without a human driver must carry 250% of the insurance usually required (*40*), although this will revert to typical coverage in 2020.

Michigan legislation has addressed many aspects of insurance. Michigan requires proof of insurance before conducting testing on a roadway without a human operator (*23*) and requires insurance for the manufacturer of the automated technology (*36*). Michigan has also addressed AV liability for upfitters when third parties provide conversions (*37*), for mechanics and auto repair providers (*36*), and for manufacturers when changes are made without their consent (*46*). Michigan also has addressed liability for auto manufacturers who have AV fleets for ridesharing in a SAVE project (*23*), specifying that the manufacturer must have $10 million in liability insurance and shall be liable when the ADS is at fault (*47*).

Operator Requirements

Operator requirements for autonomous vehicles are an important consideration and were addressed in the first legislation in Nevada in 2011, which required the DMV to establish a driver's license endorsement for AV operation, and recognized that an AV can drive itself without active intervention of a human operator (*9*). Other states that allow AV without a human driver include Arizona (*25*), California (*44*), Colorado (*13*), Florida (*42*), Georgia (*40*), Michigan (*47*), Tennessee (*41*) and Texas (*22*). Legislation in many states does not address the driver issue (e.g., Vermont [*35*], Virginia [*38*], South Carolina [*48*], North Dakota [*31*], Pennsylvania [*49*], and Utah [*50*]), and some states explicitly require a driver in the vehicle (e.g., Washington, D.C. [*51*], Washington [*26*], New York [*45*], and Massachusetts [*28*]).

Some states recognize different requirements for vehicles with human operators present in the vehicle and for those without human operators present in the vehicle (e.g., Washington [*26*]), some specify that the ADS is considered to be licensed to operate the vehicle (e.g., Texas [*22*]), and some have different requirements based on the level of automation. Tennessee designates the operator as the person in actual physical control for a conventionally operated vehicle or the person in control of the lead vehicle in a platoon (*33*), and specifies that the ADS shall be the driver for when the ADS is engaged (*41*).

In many cases, operator requirements have changed over time. In 2012, California required that the driver must be sitting in the driving seat and monitoring the system in case of failure of system or emergency (*20*). In 2016, California legislation (*44*) waived this rule and provided authorization for tests of autonomous vehicles without a driver, steering wheel, brake pedal, or accelerator. California's 2016 legislation also recognized the role of a remote operator and required two-way communications at all times between the vehicle and the remote operator. In 2017, the California DMV published revised regulations for driverless testing and deployment (*52*), reflecting advancements in driverless technologies and the need to meet this emerging market.

Michigan also allowed operation without a person in the vehicle in 2016 (*47*), specifying that an ADS or any remote or expert-controlled assist activity shall be considered the driver or operator of the vehicle when engaged. This designation is relevant for conformance to traffic laws, and clarifies that it can electronically satisfy all physical acts required by a driver or operator of the vehicle.

In some cases, the definition of operator has implications beyond merely controlling the vehicle. In 2011, Nevada deemed that people in autonomous vehicles are not considered operators when it comes to restrictions on cell phone usage in a vehicle (*53*). Other legislation that exempts restrictions related to cell phones and other potentially distracting devices includes that of Virginia (*38*), Tennessee (*54*) and Florida (*17*), which waive limits on TV displays, integrated electronic displays, and electronic displays used in conjunction with truck platooning technology, respectively. Michigan waives cell phone use restrictions for people using SAVE, an on-demand AV taxi (*23*).

As technologies progress and driverless vehicles are increasingly the focus of testing and deployment, more states have included language that differentiates operator requirements for a human operator. For example, Colorado refers to a human operator as a natural person in the vehicle with access to steering, braking, and acceleration, and refers to an ADS as the hardware and software that can perform all aspects of the driving task (at L4 and L5) without intervention or supervision by a human operator. Colorado further clarifies that provisions that regulate a human driver and the physical driving of a vehicle do not apply to an ADS (*13*).

Texas law (*22*) states that the owner of the ADS is considered the operator for the purpose of assessing compliance with traffic laws and motor vehicle laws. This is applicable when an ADS is engaged regardless of whether the person is physically present in the vehicle.

Georgia (*40*) defines an operator as a person who drives, is in actual physical control, or causes a fully autonomous vehicle to travel with the ADS engaged. This legislation also exempts a fully autonomous vehicle or the operator of a fully autonomous vehicle from driver license requirements, if the ADS is engaged.

In many cases, the operator requirements for testing mandate that the operator be an employee or agent of the manufacturer, which is one way FAVP (*4*) differentiates testing from operation. Michigan legislation illustrates this concept and requires that employees, contractors, or other designees of the ADS manufacturer operate the AV, with exceptions for university researchers or DOT employees (*23*).

Infrastructure

Adequate infrastructure is important for the safe operation of all vehicles, including AV, but it is rarely addressed by legislation. Infrastructure needs may vary depending on the AV application: AV may depend on high-contrast lane markings, or connected vehicles may require real-time information that relies on fiber-optic cables in designated corridors and 4G traffic signals. This is reflected in Pennsylvania's legislation (*49*), which considers autonomous and connected vehicle–related technologies as allowable expenses eligible for $40,000,000 designated for municipal traffic signals.

Other legislation identifies the need to plan for AV infrastructure. Florida legislation (*42*) requires the statewide Strategic Intermodal System Plan and all long-range transportation plans developed by Florida Metropolitan Planning Organization (MPOs) to include infrastructure and technology improvements needed for AV. The Wisconsin Executive Order (*27*) mandates that an appointed committee identify roads that should be designated special corridors for AV and CV testing and operation.

In other cases, state agencies are changing their infrastructure without legislative mandate. For example, California DOT is changing lane markings from 4 in. wide to 6 in. wide to increase conspicuity for AV as well as human drivers (*55*). California is also moving away from raised pavement makers after decades of use because AV have a hard time following them (*55*). Another perspective is that AV will not require special infrastructure, because any requirement for special infrastructure significantly limits the commercial viability and potential market for a product.

Vehicle Testing and Operation

The FAVP differentiates HAV testing from operation based on whether manufacturers or their designees are operating the vehicles, and whether the resulting data is being used for analysis (*4*). However, with respect to both legislative intent and practice, in some cases it is difficult to differentiate between vehicle testing and operation.

Testing

Vehicle testing has reportedly been highest in Arizona (including Phoenix, Scottsdale, Tempe and Chandler), Pennsylvania (Pittsburg), Michigan (Detroit) and California (San Francisco) (*56*). Many states have explicitly acknowledged the intent to facilitate vehicle testing. The first legislation in Nevada provided authorization for testing and operation (*9*). Similarly, early legislation in Florida (*19*) and Michigan (*36*) authorized AV testing. In both cases, subsequent legislation changed or clarified requirements. Florida (*42*) removed requirements associated with testing, and Michigan clarified requirements with respect to insurance and operators.

Some states require specific authorization for testing. California was a lead state in early AV testing activities because it is home to many technology firms. The need for authorization and mandatory reporting has allegedly caused many firms to test vehicles in other states. In addition to permits, California requires detailed reports on vehicle disengagement and accidents or incidents, which are then made publicly available. Thirty-six companies have obtained permits to test AV in California, and in 2016 Google reported AV driving 635,868 mi (*52*).

Washington requires that testing plans be submitted to the Department of Licensing (DOL) 60 days before test, the vehicle pass an inspection, there be a U.S. licensed driver in the vehicle, compliance with all Washington laws, and proof of financial responsibility (*26*).

Not all states require authorization for testing. Testing in Arizona requires only a standard vehicle registration (*57*);

testing in Colorado and Texas requires approval only if the ADS does not comply with state and federal law. In Colorado, systems that are not compliant with current regulations need to be approved by the Colorado State Patrol and the Colorado Department of Transportation. In New York, legislation requires that the Commissioner of Motor Vehicles approve AV tests, which must be supervised by the State Police (*45*), who currently provide a police escort for all AV testing in New York (*57*).

In some cases, testing is limited to an operational domain or technology. Arkansas legislation regulates AV testing for driver-assistive truck platooning technology (*16*), Utah legislation authorized a connected vehicle technology testing program (*34*), and Connecticut prohibits AV testing for a fully autonomous vehicle on any limited access highway (*14*). Massachusetts requires a memorandum of understanding that may include speed limits and confinement to certain locations and times of day, as well as information about the vehicle, safety and previous testing, and insurance (*28*).

Operation

The framework for operation is often analogous to the framework to test AV, and in some states AV can be operated as long as they meet all federal requirements and state requirements (e.g., Florida [*42*], Texas [*22*], and Colorado [*13*]). In many states, legislation specifies requirements for license, registration, and insurance requirements, and sometimes a human operator, as previously discussed. For example, Texas does not require a licensed human operator for operation, but does require registration, title, and liability insurance, as well as a recording device (*12*).

With respect to operational considerations, Florida permits operation of an AV if the AV alerts the operator of system failure and if the operator can take control of the vehicle or assure that the automated system will safely bring the vehicle to a complete stop (*42*). This is referred to as minimal risk condition, which is included in the 15-point vehicle safety assessment (*4*) and which Georgia (*15*) defines as follows:

> a low-risk operating mode in which in which a fully autonomous vehicle operating without a human driver achieves a reasonably safe state, such as bringing the vehicle to a complete stop, upon experiencing a failure of the vehicle's automated driving system that renders the vehicle unable to perform the entire dynamic driving task.

Georgia law also specifies that no rules or regulations relative to the operation of autonomous vehicles or ADS shall limit the authority to operate such vehicles or systems.

Michigan outlines a framework for AV operation (*23*, *37*, *46*, *58*) that allows operation of an AV without a human operator in the vehicle, specifies that the ADS is the operator when the system is engaged, and designates the SAVE program, which allows operation of on-demand AV taxis by motor vehicle manufacturers. This supports Michigan's auto industry, but was opposed by technology companies such as Google and Uber (*24*). Nevada also has legislation addressing the operation of AV network companies (*12*).

Commercial Vehicle Operation

AV applications for commercial vehicles may be analogous to those for passenger cars. The most active area for CVO legislation is to allow platoons with reduced headway between vehicles. Platoons with connected vehicles were demonstrated in Europe in March 2016 (*59*), and provide benefits such as reduced fuel consumption and reduced emissions.

Supporting legislation has been passed in five states in 2017 (Arkansas [*15*], Georgia [*16*], South Carolina [*48*], Tennessee [*54*], and Texas [*18*]) and has also been incorporated into broader AV legislation (e.g., Nevada [*12*]). Typically, the legislation exempts following distance requirements for vehicles traveling in a platoon, and may not be required in states that specify a reasonable and prudent following distance. Some states may also have different rules for heavy vehicles (e.g., Michigan [*23*]), and many states provide the DOT the opportunity to deny requests for platoon operation (e.g., Arkansas [*16*], Michigan [*23*], Tennessee [*33*]) based on the plan for platoon operation that must be filed and reviewed within 30 (Tennessee and Michigan) or 45 days (Arkansas).

Privacy

Vehicle and User Data

Automated and connected vehicles generate extensive data, including location data, vehicle data (e.g., acceleration, braking, and speed data), and information about the environment (e.g., other vehicles and pedestrians). AV ride sharing services may also collect data linked with specific users, including travel data. Personal data collected during travel (e.g., search engines, emails and businesses data accessed via devices linked with the vehicle) may also be vulnerable to the extent that personal devices are linked with the vehicle and thus accessible.

Vehicle data may be of interest in case of an incident or crash, and may also be useful to document infrastructure conditions. A protocol that provides access to accident data when needed (including access and storage requirements), protects user privacy, and assures protection of proprietary components of software, is important.

Several states have developed legislation to address vehicle data and user privacy. In Michigan and Nevada, legislation addresses privacy and data handling practices for passengers of AV taxi services (e.g., Michigan [*47*], Nevada [*12*]). In California, the initial legislation required that the

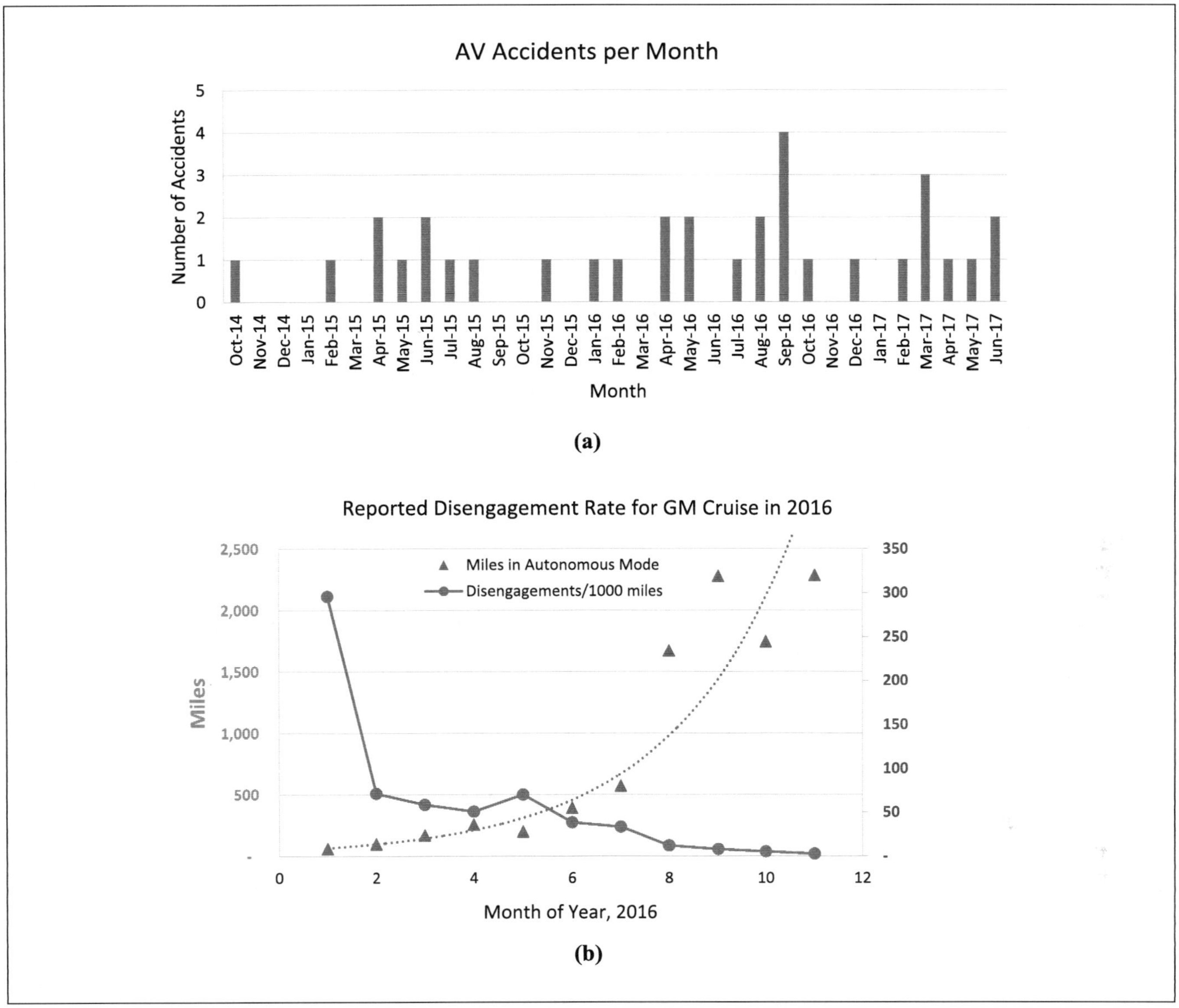

Figure 2. AV data reported in California, showing (*a*) accident data reported, and (*b*) sample disengagement data reported.
Source: (*52*).

manufacturer must provide notification of data collected to the purchaser (*20*), and subsequent legislation required disclosure of what personal information is collected (*44*).

Accident Reporting and Safety

Numerous states require that accident data be stored and retained (e.g., Nevada [*12*], California [*20*]). California requires data be collected 30 s before an accident, and retained for 3 years. California also makes information more widely available, publishing both accident and disengagement data. Sample data reported in California is shown in Figure 2. Disengagement data may reflect both manual and automatic disengagements, including unplanned and planned events.

Even states that have not promulgated data or privacy legislation have recognized that data privacy is an issue, as evidenced by North Dakota's legislation which requires a study that includes data information and storage (*31*). Texas requires that AV follow the same accident procedures as other vehicles (*22*). Georgia requires a fully autonomous vehicle involved in an accident to remain on the scene and promptly contact law enforcement (*40*); Georgia also specifies that occupants of an AV are responsible for compliance with laws relating to safety belts and child passenger restraints (*40*).

Related to safety, AV may present special concerns for emergency responders. This has been recognized by states such as Vermont that have required a report to address emergency response practices (*35*) and by states such as California that require an interaction plan be provided to assure that responders have the information needed to safely interact with an AV (*44*).

Other Considerations

Other considerations that may affect state legislation include local laws and ordinances and the recognition that the AV environment is very dynamic.

Local Laws

As noted by the FAVP, vehicles must operate within constraints of local laws as well as state laws. Local jurisdictions typically have the right to regulate vehicle operations through speed limits and other motor vehicle laws. State legislation related to AV has addressed this topic with a range of philosophies, reflecting different perspectives on the role of state and local authority.

At one end of the spectrum, Massachusetts requires the consent of local jurisdictions for AV testing and operation, as evidenced by a memorandum of understanding (MOU) (*28*). California requires that local jurisdictions provide approval for testing AV without driver (*44*).

At the other end of the spectrum, Texas (*22*), Colorado (*13*) and Illinois (*60*) explicitly state that AV regulation is a statewide concern, and local jurisdictions cannot adopt policies that would set standards other than those adopted by the state. Texas also prohibits state agencies from imposing additional AV regulations (*22*). Tennessee (*61*) prohibits local governments from banning AV. Illinois (*60*) clarifies that local governments are not prohibited from regulating ADS equipped vehicles for traffic control purposes.

Dynamic Environment

State laws may be intended to facilitate AV and associated economic development; however, state legislation is just one of many factors that affects AV activity. In many cases, legislation may not be necessary for AV to operate legally, and it may not be the catalyst to spur AV activities. As noted by one technology firm, "It is the lack of regulations that really makes Arizona attractive. In the U.S., if it isn't illegal, it's legal. Arizona hasn't passed laws on the subject—that actually makes it easier to operate there." (*56*).

Many states use legislation to signal industry that they welcome AV and associated economic development. This approach has met with mixed success. Arizona's executive order (*25*) supported several successful initiatives in a variety of cities. Michigan's legislation and test track supported the activities of their legacy auto manufacturing industry. California's legislation provided a framework for extensive testing, although there are indications that some companies have now chosen to expand testing in other states with less burdensome requirements (*62*).

Pennsylvania has limited AV legislation; however, Pittsburg has been the site for extensive AV testing, partly because of partnerships with Carnegie Melon and related technology firms. Similarly, Ohio does not have AV legislation, but has extensive AV activity, and substantial AV investments by both the public and private sector. Tennessee and Georgia have AV legislation without constraints, but have not realized significant testing or operation.

Quantifying AV activity and its impact can be challenging because most states do not have reporting requirements, and unless companies wish to leverage AV testing for publicity, proprietary considerations may warrant discretion. This may be why companies announce partnerships and programs in cities, but rarely publish details about how testing efforts are distributed across different locations. For example, Uber reported testing 150 AV in Pittsburg, San Francisco, and Tempe, and GM reported adding 130 Bolt AV to their fleet of 50 AV in Scottsdale, San Francisco, and Michigan, but neither company specified how many vehicles are in each location (*56*).

Testing may also be affected by federal activities. The U.S. DOT identified AV testing sites in the following nine states: Pennsylvania (Pittsburg), Texas (Proving Grounds Partnership in Arlington), Maryland (Aberdeen), Michigan (Willow Run, near Ypsilanti), California (Contra Costa and San Diego), Iowa (Iowa City), Wisconsin (Madison), Florida (Central Florida Partners include Orlando, the Central Florida Expressway, and Kennedy Space Center), and North Carolina (Turnpike) (*63*). The majority of these states (Iowa, Maryland, North Carolina, Texas, and Wisconsin) did not have legislation at the time of the award, and two of these states still do not have AV legislation (Maryland and Iowa), illustrating that state legislation is not necessary for testing activities.

Similarly, Columbus, Ohio, was designated a Smart City, and Ohio has several AV initiatives underway, although Ohio does not have AV legislation. Columbus reportedly used $40 million from the U.S. DOT and $10 million from industry partner Vulcan as a catalyst for $500 million in investments (as of May 2017), with aspirations for $1 billion from the private sector within 4 years (*64*). Ohio is also home to the Smart Mobility Advanced Research and Test (SMART) Center, supported by a $45 million investment by the state and Ohio State University (OSU) (*65*); and the Smart Mobility Corridor, supported by $15 million from the state and a public private partnership that includes Dublin OH, OSU, and the Transportation Research Center (*66*).

There are clearly a variety of ways to spur AV activities, and no one solution. Rhode Island (*67*) and Portland (*68*) took a different approach and sought AV partnerships via requests for proposals. Many firms are potential partners for AV activities, with hundreds of private sector companies that play a variety of roles and support numerous technology needs. The strong market presents long-term expectations for a multitrillion dollar market, but it is unclear how quickly AV goals will be realized. Implementation expectations vary dramatically; one recent study suggests that Ford and Volvo will deploy L4 by 2021, Tesla will deploy L3 or L4 by 2018, and Google will deploy L4 by 2020 (*69*).

Potential Implementation Problem: Inconsistent State Laws

As evidenced by the information in the previous sections, states have taken a wide range of approaches to AV legislation. Driver requirements, vehicle requirements, operating requirements and reporting requirements (for both accident and operational data) vary. The wide range of requirements presents current challenges for manufacturers, and future challenges for fleet operators, human operators, and AV consumers.

State requirements for traditional vehicles and drivers have a common basic framework for licensure, registration, and operation. As a result, there is basic reciprocity for operation across state lines and jurisdictions, ensuring seamless interstate transportation. On the other hand, the requirements for AV are evolving in a variety of directions, which in some cases are even diametrically opposed. For example, some states require that a driver must be in an AV and able to take manual control (e.g., Connecticut [*12*], Washington [*26*], Massachusetts [*28*], and New York [*45*]), which contradicts other states such as North Carolina (*70*) that not only prohibit a driver from taking control in a fully autonomous vehicle, but also require that any equipment that allows an occupant to take control be removed or disabled (e.g., North Carolina [*70*]). The diversity of regulatory requirements presents obstacles to interstate commerce, and potentially even intercity commerce in states that hold home rule sovereign.

The potential problems associated with the current "patchwork of state regulations" cannot be overstated, and have been one impetus for the proposed regulations at the federal level. Federal legislation that provides consistency across state lines will ensure seamless boundaries between states and between local jurisdictions, and will also support the development of AV technologies in the U.S., rather than in other countries where state borders do not present regulatory barriers.

Conclusions

AV technology is rapidly advancing and states have responded with a variety of legislation addressing traditional topics such as licensing and registration, as well as new topics, such as AV operator requirements.

- Registration and license requirements vary. AVs may require (Georgia and Tennessee) or allow (Texas) registration that designates them as automated vehicles, and may (Michigan) or may not (Virginia) require special plates. Few states have leveraged aggressive fees on AV, probably because of the limited number of AVs, and the desire to promote AV testing. States have authorized fees to cover administration costs for AV (California), levied a 3% excise tax for AV network companies (Nevada), and have prohibited local jurisdictions from charging AV fees or taxes (Michigan and Nevada).
- Insurance requirements may be the same as for other vehicles (Texas and Colorado), may be higher (250% of the standard coverage in Georgia), or may vary depending on the application ($5 milion for testing in Nevada, California, and Connecticut, and $10 million for AV rideshare fleets in Michigan).
- Operator requirements differ from state to state and may change over time within a given state. Some states allow an AV without a driver (Arizona, California, Colorado, Florida, Georgia, Michigan, Tennessee and Texas) and other states explicitly require a driver in the vehicle who is able to provide override capabilities, if needed (Washington, Massachusetts, New York, Connecticut, and Washington D.C.). Some states have recognized that the operator may be remotely located (California), in which case two-way communications are required.

States have a variety of AV legislation, and similar legislation has resulted in different outcomes for AV activities. This illustrates that legislation is one of many factors considered when companies decide where to test and deploy their vehicles. The breadth of legislation and executive orders provides a wealth of information for other states to learn from when charting a path for the future. An excellent resource for the most current information about legislation is the NCSL website: http://www.ncsl.org/research/transportation/autonomous-vehicles-self-driving-vehicles-enacted-legislation.aspx, which includes an overview of federal activities and a summary of state legislation.

The wide range of legislative approaches has resulted in a "patchwork" of laws that make it difficult for manufacturers and present an obstacle to interstate transportation. To support interstate commerce and promote AV development in the U.S., both the Senate and House have proposed legislation to expand the role of the federal government beyond safety to HAV operation, and increase the number of AVs sold under FMVSS exemptions from 2,500 to 100,000 per year per manufacturer, with incremental increases over 3 years.

Acknowledgments

This work was supported in part by the Joint Transportation Research Program administered by the Indiana Department of Transportation and Purdue University. The author would also like to thank Mike Smith, Jim Sturdevant and Tim Wells of INDOT, and Keith Hoernschemeyer and Karen Stippich of FHWA for serving on the Advisory Committee and providing valuable input. The author would also like to thank Amanda Essex of the NCSL, the Indiana Motor Trucking Association, and Tony Petraglia for their input.

References

1. Clements, L., and K. Kockelman. Economic Effects of Automated Vehicles. Presented at 96th Annual Meeting of the Transportation Research Board, Washington, D.C., 2017.
2. Morris, D. Driverless Cars Will Be Part of a $7 Trillion Market by 2050. *Fortune*. June 3, 2017.
3. Fagnant, D., and K. Kockelman. Preparing a Nation for Autonomous Vehicles: Opportunities, Barriers and Policy Recommendations. *Transportation Research Part A: Policy and Practice*, Vol. 77, 2015, pp. 167–181.
4. National Highway Traffic Safety Administration. *Federal Automated Vehicles Policy: Accelerating the Next Revolution in Roadway Safety*. U.S. Department of Transportation, Washington, D.C., 2016.
5. National Highway Traffic Safety Administration. US DOT Releases New Automated Driving Systems Guidance. US DOT. https://www.nhtsa.gov/press-releases/us-dot-releases-new-automated-driving-systems-guidance. Accessed November 5, 2017.
6. Naughton, K. Ford's Dozing Engineers Side With Google in Full Autonomy Push. *Bloomberg Technology*. Bloomberg, LP. 2017.
7. United States HR 3388. *Safely Ensuring Lives Future Deployment and Research in Vehicle Evolution Act or the SELF DRIVE Act*. 2017.
8. United States S 1885. *American Vision for Safer Transportation through Advancement of Revolutionary Technologies Act or the AV START Act*. United States Congress, 2017.
9. Nevada AB 511. *An Act Relating to Transportation; Authorizing in This State the Operation of, and a Driver's License Endorsement for Operators of Autonomous Vehicles*. Nevada Legislature, 2011.
10. National Conference of State Legislatures (NCSL). *Autonomous Vehicles, Self-Driving Vehicles Enacted Legislation*. NCSL, October 23, 2017. http://www.ncsl.org/research/transportation/autonomous-vehicles-self-driving-vehicles-enacted-legislation.aspx. Accessed November 5, 2017.
11. Louisiana HB 1143. *An Act To enact R.S. 32:1(1.2), Relative to Motor Vehicles; to Define "Autonomous Technology" for Purposes of Highway Regulatory Provisions; and to Provide for Related Matters*. Legislature of Louisiana, 2016.
12. Nevada AB 69. *An Act Relating to Transportation; Revising Requirements for the Testing or Operation of an Autonomous Vehicle on a Highway within This State*. Legislature of Louisiana, 2017.
13. Colorado SB 213. *Concerning Authorization for Automated Driving Systems to Control Motor Vehicles throughout Colorado*. Colorado General Assembly. Southern California Public Radio, 2017.
14. Connecticut SB 260. *An Act Concerning Autonomous Vehicles*. Connecticut General Assembly, 2017.
15. Georgia HB 472. *An Act to Amend Title 40 of the Official Code of Georgia Annotated, Relating to Motor Vehicles and Traffic, so as to Provide for an Exception for Following Requirements for Non-Leading Vehicles Following in a Coordinated Platoon*. General Assembly of Georgia, 2017.
16. Arkansas HB 1754. *An Act to Regulation the Testing of Vehicles with Autonomous Technology, and for Other Purposes*. Arkansas General Assembly, 2017.
17. Florida HB 7061. *An Act Relating to Transportation*. Legislature of the US State of Florida, 2016.
18. Texas HB 1791. *An Act Relating to the Use of Connected Braking Systems to Maintain Distance between Vehicles*. Texas Legislature, 2017.
19. Florida HB 1207. *An Act Related to Vehicles with Autonomous Technology. Autonomous Technology, and for Other Purposes*. Legislature of the US State of Florida, 2012.
20. California SB 1298. *Autonomous Vehicles: Safety and Performance Requirements*. California State Legislature, 2012.
21. Tennessee SB 1561. *An Act to Amend Tennessee Code Annotated, Title 47; Title 54, Chapter 1; Title 55 and Title 67, Relative to Autonomous Vehicles*. Tennessee General Assembly, 2016.
22. Texas HB 2205. *An Act Relating to Automated Motor Vehicles*. Texas Legislature, 2017.
23. Michigan SB 995. *An Act to Amend 1949 PA 30*. Michigan Legislature, 2016.
24. Korosec, K. Michigan Just Passed the Most Permissive Self-Driving Car Laws in the Country. *Fortune*. December 9. Time Inc., 2016.
25. Ducey, D. *Self-Driving Vehicle Testing and Piloting in the State of Arizona; Self-Driving Vehicle Oversight Committee*. Arizona Executive Order 2015-09. Office of Arizona Governor, 2015.
26. Inslee, J. *Autonomous Vehicle Testing and Technology*. Washington Executive Order 17-02. Governor of State of Washington, 2017.
27. Walker, S. *Relating to the Creation of the Governor's Steering Committee on Autonomous and Connected Vehicle Testing and Deployment*. Wisconsin Executive Order #245. State of Wisconsin Office of the Governor, 2017.
28. Baker, C. *To Promote the Testing and Deployment of Highly Automated Driving Technologies*. Massachusetts Executive Order 572. Governor of Massachusetts, 2016.
29. Carney, J. *Establishment of the Advisory Council on Connected and Autonomous Vehicles*. Delaware Executive Order Number 14. Governor of Delaware, 2017.
30. Alabama SJR 81. *Establishing the Joint Legislative Committee to Study Self-Driving Vehicles*. Alabama Legislature, 2016.
31. North Dakota HB 1202. *An Act to Provide for a Department of Transportation Study*. North Dakota Legislative Assembly, 2017.
32. North Dakota HB 1065. *An Act to Provide for a Legislative Management Study of Automated Motor Vehicles*. North Dakota Legislative Assembly, 2015.
33. Tennessee SB 676. *An Act to Amend Tennessee Code Annotated, Title 5; Title 6; Title 7; Title 39; Title 40; Title 54; Title 55; Title 56; Title 65 and Title 67, Relative to Vehicles Operated by Unconventional Means*. Tennessee General Assembly, 2017.
34. Utah HB 373. *Connected Vehicle Testing*. Utah State Legislature, 2015.
35. Vermont HB 494. *An Act Relating to the Transportation Program and Miscellaneous Changes to Transportation Related Law*. Vermont General Assembly, 2017.
36. Michigan SB 169. *An Act to Amend 1949 PA 300*. Michigan Legislature, 2013.

37. Michigan SB 663. *An Act to Amend 1961 PA 236*. Michigan Legislature, 2013.
38. Virginia HB 454. *An Act to Amend and Reenact Sections 46.2-750 and 46.2-1077 of the Code of Virginia, Relating to Motor Vehicles Equipped with Televisions and Video; Not within View of Driver; License Plates on Vehicles Owned by the Commonwealth*. Virginia General Assembly, 2016.
39. Buchanan-King, M. Virginia Tech Transportation Institute Part of New State Legislation Enacted for the Testing and Use of Automated-Vehicle Technologies. *Virginia Tech News*. June 11. Virginia Polytechnic Institute and State University, 2016.
40. Georgia HB 219. *An Act to Amend Title 40 of the Official Code of Georgia Annotated, Relating to Motor Vehicles and Traffic, so as to Provide for Definitions; to Exempt Persons Operating a Fully Autonomous Motor Vehicle with the Automated Driving System Engaged from the Requirement to Hold a Driver's License*. Virginia Polytechnic Institute and State University, 2017.
41. Tennessee SB 151. *An Act to Amend Tennessee Code Annotated, Title 5; Title 6; Title 7; Title 39; Title 40; Title 54; Title 55; Title 56; Title 65 and Title 67, Relative to Autonomous Vehicles*. Tennessee General Assembly, 2017.
42. Florida HB 7027. *An Act Relating to the Department of Transportation*. Legislature of the U.S. State of Florida, 2016.
43. Nevada SB 313. *An Act Relating to Autonomous Vehicles*. Nevada Legislature, 2013.
44. California AB 1592. *Autonomous Vehicles: Pilot Project*. California State Legislature, 2016.
45. New York. SB 2005. *A Budget Bill*. New York State Legislature, 2017.
46. Michigan SB 997. *An Act to Amend 1949 PA 30*. New York State Legislature, 2016.
47. Michigan SB 996. *An Act to Amend 1949 PA 30*. Michigan Legislature, 2016.
48. South Carolina HB 3289. *A Bill to Amend Section 56-5-1930, Code of Laws of South Carolina, 1976, Relating to the Distance That Must Be Maintained between Vehicles Traveling along a Highway*. South Carolina General Assembly, 2017.
49. Pennsylvania SB 1267. *Transportation Code (74 PA.C.S) and Vehicle Code (75 PA.C.S)—Omnibus Amendments*. Commonwealth of Pennsylvania, 2016.
50. Utah HB 280. *Study Related to Autonomous Vehicles*. Utah State Legislature, 2016.
51. Washington D.C. 2012 DC B 10931. *To Authorize Autonomous Vehicles to Operate on District Roadways, to Require the Department of Motor Vehicles to Create an Autonomous Vehicle*. Council of the District of Columbia, 2012.
52. California Department of Motor Vehicles. *Autonomous Vehicles in California*. State of California Department of Motor Vehicles, 2017. https://www.dmv.ca.gov/portal/dmv/detail/vr/autonomous/bkgd. Accessed July 14, 2017.
53. Nevada SB 140. *An Act Relating to Traffic Laws; Prohibiting a Person from Using a Cellular Telephone or Other Handheld Wireless Communications Device while Operating a Motor Vehicle in Certain Circumstances*. Nevada Legislature, 2011.
54. Tennessee SB 2333. *An Act to Amend Tennessee Code Annotated, Title 55, Chapter 8 and Title 55, Chapter 9, Part 1, Relative to Motor Vehicles Equipped with Autonomous Technology*. Tennessee General Assembly, 2016.
55. Carpenter, S. California is Changing Its Roads for Self-Driving Cars. *Take Two*. Southern California Public Radio, July 12, 2017. http://www.scpr.org/programs/take-two/2017/07/12/57901/california-is-changing-its-roads-for-self-driving/. Accessed November 5, 2017.
56. Burden, M. Phoenix Emerging as Michigan Rival for Self-Driving Tests. *Detroit News*. Digital First Media, July 3, 2017.
57. Marshall, A. Congress Finally Gets Going on That Regulating Robocars Thing. *Wired*. Condé Nast International, June 21, 2017.
58. Michigan SB 998. *An Act to Amend 1961 PA 236*. Michigan Legislature, 2016.
59. Hirsch, J. Daimler Tests Self-Driving Truck Platoon in Live Traffic. *Trucks.com*. Trucks.com International, Inc., March 21, 2016. https://www.trucks.com/2016/03/21/daimler-tests-self-driving-truck-platoon-in-live-traffic/. Accessed November 5, 2017.
60. Illinois HB 791. *An Act Concerning Transportation*. Public Act 100-0352. Illinois General Assembly, 2017.
61. Tennessee SB 598. *An Act to Amend Tennessee Code Annotated, Title 4; Title 5; Title 6; Title 7 and Title 55, Relative to the Regulation of Autonomous Vehicles*. Tennessee General Assembly, 2015.
62. Gear, A. California's New Self-Driving Car Rules Are Great for Texas. *Wired*. Condé Nast International, December 17, 2016.
63. U.S. Department of Transportation. U.S. Department of Transportation Designates 10 Automated Vehicle Proving Grounds to Encourage Testing of New Technologies. *Briefing Room*. US DOT, January 19, 2017.
64. Maddox, T. How Columbus, Ohio, Parlayed $50 Million into $500 Million for a Smart City Transportation Network. *TechRepublic*. CBS Interactive, May 10, 2017.
65. IndustryWeek. *Ohio, OSU to Invest $45 Million in Autonomous Vehicle Testing Expansion*. Informa USA, Inc., January 26, 2017.
66. Johnson, R. Self-Driving Vehicle Testing Expands on Ohio's Smart Mobility Corridor via New Partnership. *Statescoop*. Scoop News Group, June 12, 2017.
67. Rhode Island. *Request for Information: Connected and Autonomous Vehicles and Other Innovative Transport System Technologies Framework for Implementation and Integration*. State of Rhode Island and Providence Plantations, June 7, 2017.
68. Portland, Oregon. Smart Autonomous Vehicle Initiative. *Request for Information Number 00000657*. City of Portland, June 9, 2017.
69. Lewis, P., G. Rogers, and S. Turner. *Beyond Speculation Automated Vehicles and Public Policy*. Eno Center for Transportation, Eno Transportation Foundation, 2017.
70. North Carolina BH 469. *An Act to Regulate the Operation of Fully Autonomous Motor Vehicles on the Public Highways of this State*. North Carolina General Assembly, 2017.

The Standing Committee on Strategic Management (ABC10) peer-reviewed this paper (18-00586).

The contents of this paper reflect the views of the author, who is responsible for the facts and the accuracy of the data presented here, and do not necessarily reflect the official views or policies of the sponsoring organizations. These contents do not constitute a standard, specification, or regulation.

TRR
JOURNAL OF THE TRANSPORTATION RESEARCH BOARD

Transportation Research Record
2018, Vol. 2672(7) 14–23
© National Academy of Sciences:
Transportation Research Board 2018
Article reuse guidelines:
sagepub.com/journals-permissions
DOI: 10.1177/0361198118796392
journals.sagepub.com/home/trr
SAGE

Who Will Own and Have Propriety over Our Automated Future? Considering Governance of Ownership to Maximize Access, Efficiency, and Equity in Cities

Hamish Campbell[1]

Abstract

Automated vehicle technology presents an opportunity to remake urban mobility in a way that maximizes access, efficiency, and equity. One of the roles for policymakers is to ensure that future governance of automated vehicles (AVs) achieves this. When considering governance, the current literature centers on issues related to the safe operation and deployment of AVs but has not fully considered the implications of AV ownership and ridesourcing platform data propriety on achieving the most desirable urban mobility outcomes. Specifically, the literature has not considered: a future scenario where individually owned AVs are shared when not in use; and the implications of ridesourcing platform data remaining proprietary in future. This paper analyzes why: the future of AV ownership may not be a binary choice between owning an AV/not sharing and sharing an unowned fleet, which is the current consensus in the literature; the incentives for consumers to simultaneously own an AV and share it when they are not using it could be high; the way ridesourcing platform data is collected, used, and shared could be a very influential factor for urban mobility outcomes, but its implications have not been robustly analyzed in the literature; and future scenario-building and modeling should consider the implications of widespread sharing of individually owned AVs, as well as the implications of ridesourcing platform data propriety on urban mobility outcomes. Developing a foundation for future good governance of AV ownership and ridesourcing platform data propriety should be an immediate priority for researchers, policymakers, and practitioners.

Anthony Foxx, the former Secretary of the U.S. Department of Transportation, stated at the 2016 Automated Vehicles Symposium that the choice for "government agencies is to act or react" to the emergence of AV technology. What he meant by this is that policymakers need to start preparing the transportation ecosystem for the integration of these new types of vehicles (*1*).

AV technology promises the possibility of two major overarching benefits for cities:

1. an opportunity to substantially improve the safety and operation of the transportation system; and
2. an opportunity to remake mobility in way that maximizes access, efficiency, and equity.

Given the magnitude and complexity of both of these opportunities, the current race towards vehicle automation has been likened to the 1960s race to the moon. There is general agreement about where we want to go, but nobody's sure how we're going to get there.

The role for policymakers is to ensure an appropriate framework is in place for good governance of AVs. The United Nations Educational, Scientific and Cultural Organization (UNESCO) defines governance as the structures and processes that are designed to ensure accountability, transparency, responsiveness, rule of law, stability, equity and inclusiveness, empowerment, and broad-based participation (*2*).

To date, much of the discussion in the literature about AV governance centers on issues of safety, and what the regulatory and legislative roles and responsibilities between federal, state, and municipal governments should be to promote safe operation, testing, certification, and deployment. Debates have centered on questions such as: does the robot driving the car require

[1]City of Mississauga, Mississauga, ON, Canada

Corresponding Author:
Address correspondence to Hamish Campbell:
hamish.campbell@mississauga.ca

licensing the same way a human driver does, and who is liable in circumstances where the software fails? Others have expressed concern that humans will increasingly trust self-driving vehicle technology too quickly, as Joshua Brown did in 2016 when his Tesla collided with a semi-truck trailer that the vehicle's autopilot feature did not detect (*3*). There is also the famous "trolley dilemma" which asks how an AV should be programmed to react when it has no choice but to cause harm. All of these questions are important, and their answers are complex. To mitigate the stifling of innovation on AV deployment, there is an overarching consensus that a balance must be maintained between a regulatory regime that ensures ongoing safety of the traveling public but that also doesn't hamper progress towards AV deployment.

This focus on safe operation and deployment of AVs makes sense. It still remains unclear when, where, and at what level of automation AVs will reach scale on our streets. There is widespread disagreement about when the technology will be sufficiently mature to handle the rigorous complexities of a typical vehicle trip, such as obeying all local traffic laws, understanding four-way stops at intersections, negotiating right-turn-on-red situations, handling worn-out lane markings or unpaved roads, and recognizing all other road users

Notwithstanding these hurdles, this paper assumes that safe, full (Level 5) vehicle automation is achievable in the short to medium term (5–10 years). The author recognizes that there are many technical barriers that will need to be overcome to achieve this level of automation in such a short timeframe; however, given that many automakers and technology companies have promised that the technology is nearing a stage where it is able to safely handle nearly all traffic situations and weather, this timeline needs to be taken seriously.

This leaves policymakers to consider what is required for governance of both the hardware (i.e., AVs) and the software (i.e., ridesourcing platforms) in a way that safeguards access, efficiency, and equity in an automated future. The current literature offers few answers. There is an understanding of what is possible; Janette Sadik Khan, former New York City Department of Transportation (DOT) Commissioner, transportation principal at Bloomberg Associates, and National Association of City Transportation Officials (NACTO) Chair captures it well: "Autonomous vehicles offer a once-in-a-lifetime opportunity to reset our streets and address the fundamental issues of congestion, traffic safety, congestion and mobility as our cities grow ever larger." (*4*) From a governance perspective though, what do transportation policymakers need to consider, in order to achieve this envisioned outcome?

NACTO's policy paper (*5*) outlines a number of tangible strategies for policymakers to pursue regarding governance of AVs; however, these strategies center on the safe deployment and operation of AV technology and its relationship with city infrastructure. NACTO does not consider how the governance of ownership and ridesourcing data propriety could impact urban mobility outcomes.

This paper proposes that in order to shape a mobility future where access, efficiency, and equity are maximized for cities and where the most desirable urban mobility outcomes are realized, the following two possibilities will need much greater consideration by researchers, policymakers, and practitioners:

1. A future scenario where individually owned AVs are shared when not in use; and
2. The implications of ridesourcing platform data remaining proprietary in future.

More precisely, the following are questions regarding these possibilities that will need further consideration:

1. Will any individual and/or entity be allowed to own and/or share an AV and/or fleet of AVs in future? What is the right approach for good governance of AV ownership to achieve the most desirable urban mobility outcomes?
2. Will ridesourcing platform data collection and use remain proprietary in future? What is the right approach for good governance of the means for vehicle sharing to achieve the most desirable urban mobility outcomes?

How AV hardware and ridesourcing software are owned and data is shared will be very influential for cities; yet, this paper has found that the literature yields few answers to these questions. If the overarching goal is to ensure a future where access, efficiency, and equity are maximized for cities, as well as reduce the inefficiencies of today's mobility paradigm, this paper will explore why these questions are of importance; and, why:

- the future of AV ownership may not be a binary choice between owning an AV/not sharing and sharing an unowned fleet, which is the current consensus in the literature;
- the incentives for consumers to simultaneously own an AV and share it when they are not using it could be high;
- the way ridesourcing platform data is collected, used, and shared could be a very influential factor for urban mobility outcomes, but its implications have not been robustly analyzed in the literature; and

- future scenario-building and modeling should consider the implications of widespread sharing of individually owned AVs, as well as the implications of ridesourcing platform data propriety on urban mobility outcomes.

Electric, Automated, Shared, Connected, and Mobility as a Service

Before exploring governance of AV ownership and ridesourcing platform data propriety further, it is important to first provide an overview of what is generally agreed to be the most desirable future urban mobility outcome when vehicle automation reaches scale.

While there are many variations on the vision and the possible benefits, the general consensus is that the future of mobility will be: i) electric, ii) automated, iii) shared, and iv) connected. This paper uses the acronym EASC (i.e., electric, automated, shared, connected) to describe this future scenario. Along with EASC, there is also the concept of transportation/mobility as a service (this paper uses the acronym MaaS), which proposes that the new transportation paradigm more broadly is one where people will use and pay for transportation services on an as-needed basis, rather than owning them. MAASiFie (*6*) defines MaaS as "multimodal and sustainable mobility services addressing customers' transport needs by integrating planning and payment on a one-stop-shop principle."

It is important to note that EASC and MaaS are not the same thing; EASC broadly describes possibilities for individuals, groups, and companies (i.e., the micro level) while MaaS broadly describes possibilities for the overall system (i.e., the macro level). If and when the convergence of EASC and MaaS is achieved, the future of urban mobility could be one where vehicle ownership is substantially reduced. In its place, shared fleets of automated electric vehicles will drive around the streets of our cities and suburbs, picking up and dropping off passengers on an as-needed basis. These on-demand vehicles will be complemented by high-quality transit, bikeshare and pedestrian-oriented streets. As Robin Chase, cofounder of Zipcar writes, the future "is not the sale of cars, it is the sale of trips: transportation as a service." (*7*)

The anticipated benefits of achieving a future where EASC and MaaS converge are well documented in the literature. Although each city would be unique, early modeling and research has proposed that the benefits could include:

- **Less cars and greater network efficiency** due to vehicle sharing; Qinglu, Kockelman and Segal (*8*) summarize the results of eight different modeling exercises. From Austin to Singapore to New Jersey to Lisbon; all studies have concluded that if EASC and MaaS converge as the predominant transportation model, it would result in between 6 and 11 conventional vehicles being replaced with one shared vehicle.
- **Lower greenhouse gas emissions (GHG)** from moving to an entirely electric vehicle fleet that is lighter and includes technology-enabled efficiencies, such as platooning.
- **Improved access and equity** for children, seniors, and the disabled who cannot drive.
- **More affordable housing** resulting from reduced costs to build parking.
- **Less parking** (if less cars) resulting in more land for redevelopment increasing the density of our cities.

The incredible rise of Uber and Lyft in the United States, Daimler's Mytaxi and Hailo service, nuTonomy in Singapore, and Didi Chuxing in China are early indications that sharing is popular. As the Brookings Institution (*9*) notes, Uber launched in 2010 and in 2016 reached a valuation of $68 billion; Lyft attracted a $500 million in investment from General Motors around the same time; Didi Chuxing meanwhile has been valued at $35 billion and at the time of writing, makes 14 million trips each day in 400 cities in China.

Many automakers, such as General Motors (GM), Ford and Daimler are now repositioning themselves to be mobility and technology companies for a future of EASC and MaaS convergence. For example, GM acquired Cruise, which is launching an app for its employees to hail rides from one of its vehicles in San Francisco. GM has also launched a ride-sharing service called Maven which allows some GM owners to rent their vehicles out when not in use.

Achieving the hoped-for convergence of EASC and MaaS would, however, depend on a major shift in consumer behavior (i.e., widespread sharing of vehicles and rides) as well as a major shift in the way the transportation system works (i.e., that all tripmaking would be optimized to maximize network efficiency). The next section explores these assumptions in more detail and their challenges.

Assumptions about EASC and MAAS Convergence

For EASC and MaaS to converge in a way that will maximize benefits for cities, there are four overarching assumptions about consumers' future behavior:

1. They will be willing to forgo the act of driving and hand over this duty to a robot
2. They will be willing to "give up" the internal combustion engine and embrace fully electric vehicles
3. They will be willing to share and have their family members (children, older adults) share a ride with strangers (if sharing a trip) and/or in a vehicle that strangers have ridden in (if taking a trip in a shared vehicle)
4. They will be willing to forgo vehicle ownership and purchase trips on demand

Combined, these four assumptions are fairly substantive and taken together would be a monumental shift in behavior to that of today's motoring public. The literature is largely inconclusive on how viable all of these assumptions are; however, there is some data to support that the first three assumptions may be reasonable. For the purposes of this paper, it is assumed that the first three are feasible (i.e., that people will be willing to forgo driving, "give up" the internal combustion engine for one that is electric, and share a ride and/or ride in a shared vehicle that strangers have ridden in).

The fourth assumption is less clear. Will consumers be willing to forgo vehicle ownership and instead purchase trips on demand from a shared fleet(s)? The general consensus in the literature is that because cars generally sit unused 95% of the time, the economics will be so compelling (i.e., cost per mile so low) that nearly everyone will choose to simply buy trips on an as-needed basis and therefore forgo ownership of an AV. This paper concludes that this premise could be incorrect and has found that the incentives to simultaneously own and share an AV when not in use may be much more compelling for consumers than simply buying trips on-demand (more on this below).

In addition to the above-noted assumptions about consumers' future behavior, EASC and MaaS convergence also relies on the ability to achieve optimized tripmaking across the entire transportation network. This would mean a major change to the way the business models of ridesourcing platforms, such as Uber, Lyft, and Didi, work today. Today, all ridesourcing platforms' data and algorithms are owned by the platform where they are collected and are not shared in any meaningful way with one another and/or government agencies. There is neither an overarching regulatory framework for the data that these platforms collect, nor much in the way of regulations about how the data collected by these platforms is used. Each platform's algorithms and data are proprietary. If ridesourcing platform data propriety continues, meaning ownership of data on and from these networks is protected, optimizing tripmaking across the entire transportation network to fully realize the benefits of EASC and MaaS convergence would likely be challenging.

Notwithstanding the above-noted challenges with the assumptions that underpin convergence of EASC and MaaS, all of the literature reviewed generally concludes:

- the future choice for consumers will be binary, between the individual AV ownership scenario or the shared fleet scenario;
- convergence of EASC and MaaS will emerge as the predominant mobility model because the economics will be so compelling to share; and, as a result,
- vehicle ownership will become unpopular and decrease substantially as the convenience of sharing increases resulting in less cars, less parking, greater access, etc.

The literature alludes to the need for greater data sharing, but no discussion of ridesourcing platform data propriety and its implications for urban mobility outcomes was found.

The following are excerpts from a sample of literature to show that the current consensus presents the future as a binary choice, between individual AV ownership/not sharing or not owning/using a shared fleet:

- Bloomberg New Energy Finance and McKinsey & Company (*10*) conclude "The costs of a pooled self-driving taxi could be 30–60 percent less than a private vehicle. Moreover, this model may offer additional consumer utility: the enjoyment of on-demand access to a variety of vehicles for door-to-door transport, without the upfront cost or responsibility for maintenance attached to vehicle ownership."
- Boston Consulting Group (*11*) concludes "A third scenario examines how fully autonomous vehicles could impact urban transportation methods. Here, the economics of sharing robo-taxis in cities becomes more attractive than the economics of conventional taxis and even of personal-vehicle ownership. This so-called Droogle scenario (we created the term by combining the words 'drone' and 'Google') is an aggressive projection: it assumes that the favorable economics of robo-taxis will lead many of the world's largest cities to encourage or even mandate the use of AVs for personal transportation. Under this scenario, drivers in urban and near-suburban settings would either choose to give up their personal vehicles or be compelled to do so."
- Brookings Institution (*9*) concludes "Ride-sharing alternatives or taxis will reduce congestion because

young people in many places have a preference for those options as opposed to purchasing their own vehicle. For example, a Roland Berger survey showed that '51 percent of Chinese car owners said they would prefer to use robot taxis rather than buy a new vehicle themselves, compared with 26 percent of Americans.' And in Japan, the company Robot Taxis says that 'it hopes to develop a fleet of thousands of driverless vehicles in time for Tokyo [Olympics] 2020'."

- Cohen and Cavoli (*12*) conclude "Although different roll-out and use scenarios have been described in the literature, there seems to be a broad consensus that two are likely to emerge – independently or in parallel: on the one hand, the "business-as-usual" scenario and on the other hand the "shared mobility" scenario."
- The European Parliament, Directorate-General for Internal Policies of the Union (*13*) state "This shift towards a shared, mobility service rather than a private ownership model is a big step for auto manufacturers" and conclude that "connected driverless taxi services have the potential to improve accessibility and reduce car ownership."
- Forbes (*14*) states "Car sharing is already growing in popularity. When getting a ride someplace is as easy and inexpensive as ordering an automated Uber, we can expect a significant percentage of people to realize car ownership is a thing of the past."
- Morgan Stanley (*15*) concludes "The market for private automobile ownership is likely headed for disruption."
- Qinglu, Kockelman and Segal (*8*) conclude "One might expect the early integration of SAVs to cater to the shift in urban living where vehicle ownership is the most expensive and cumbersome."
- RAND Corporation (*16*) concludes "It is quite likely that the availability of driverless taxis could likewise motivate some households to reduce levels of auto ownership."
- The Rocky Mountain Institute (*17*) states "Consumers could choose automated mobility services exclusively instead of personal vehicle ownership."
- UITP/International Association of Public Transport (*18*) concludes "If employed as shared robo-taxis and mini-buses as well as used to reduce car ownership through more effective car-sharing schemes, driverless AVs could dramatically enhance public transport."
- The U.S. DOT (*19*) concludes "Highly automated vehicles could change the way many people access mobility, increasing the use of on-demand car services, both personal and pooled, and reducing the need for personal vehicle ownership."

Figure 1 is a schematic that visualizes this current consensus. What the literature overwhelmingly concludes is that the convergence of EASC and MaaS is the recipe to maximize access, efficiency, and equity for cities and achieve the most desirable urban mobility outcomes, and that their convergence is assured because the cost per mile of using a shared fleet will be so much lower than the cost per mile of individual AV ownership. Economics will drive consumers towards the shared fleet solution.

Much of the literature also includes analysis and modeling to support this premise. As one example, Stanford economist Tony Seba and his colleagues calculated the cost per mile of individual AV ownership scenario versus the shared fleet scenario. Arbid and Seba (*20*) determined that the cost per mile of the future choices would be as follows:

1. Operate a new individually owned vehicle = cost per mile ranging from 61–78 cents by 2030
2. Operate a used individually owned vehicle = cost per mile ranging from 31–34 cents by 2030
3. Use shared fleet = cost per mile ranging from 10–16 cents for riding alone, 3–5 cents for MaaS pooling

Arbid and Seba (*20*) conclude: "From the introduction of MaaS, consumers considering the purchase of a new car will be faced with new economics, in which choosing MaaS over Individual Ownership will lead to a four- to ten-times reduction in costs. We know of no other market where a 10x cost differential has not led to a disruption. This very significant cost differential will be the key

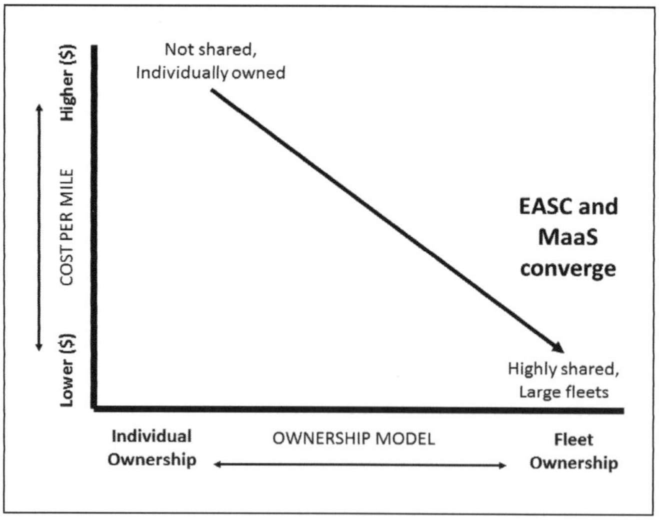

Figure 1. Current consensus—mobility model of the future.

driver for rapid and widespread MaaS adoption for car owners. Potential car buyers will stop buying new cars."

While this paper has singled out Arbid and Seba's (*20*) analysis, it is important to re-state that all of the literature comes to the same conclusion: consumers' desire to shift to a future scenario where shared fleets predominate is a rational alternative to individual AV ownership because taking trips on the shared fleet will be cheaper and, as such, the economics will push most consumers to it and away from individual AV ownership.

The potential flaw with this premise is that that ownership of an AV is not mutually exclusive to vehicle sharing. Put differently, high levels of individual AV ownership do not preclude the possibility of simultaneous widespread vehicle sharing of these individually owned AVs.

By the literature assuming that the future of AV ownership is a binary choice between owning/not sharing vs. sharing/not owning, there is a false, underlying assumption that consumers will be willing to hand over the value created by vehicle automation (i.e., the robot) to somebody else (i.e., a fleet operator, auto manufacturer and/or ridesourcing platform).

This paper finds that there could be compelling incentives for consumers to own an AV and simultaneously share it when not in use because a robot is now creating value for whomever owns the AV.

The next section of this paper explains why an additional alternative future scenario of sharing individually owned AVs requires greater consideration by researchers, policymakers, and practitioners.

Self-Interest, Value Capture and an Alternative Mobility Model for Future EASC and MAAS Convergence

Consider this question: if an individual consumer is able to purchase an AV, and that same individual can then share his/her AV when it is not in use (estimated to be 95% of the time) and monetize his/her under-used asset, why wouldn't he/she? The sharing economy has become as big as it has because it allows owners to monetize their owned assets that would otherwise sit idle.

Adam Smith, the father of modern economics said in his most famous book, *The Wealth of Nations*, that the largest motivator of economic activity is self-interest. One of his most famous quotes is: "It is not from the benevolence of the butcher, the brewer or the baker that we expect our dinner, but from their regard to their own self-interest." With automation, there is suddenly tremendous value being created by a robot. If Moore's law (which is the loose observation that computing power generally doubles and its costs fall every 18 months) ends up applying to AVs, the hardware and technology that makes electric, automated vehicles expensive today could end up costing much less in future thereby making the cost of an AV similar or lower to that of a modest car today. This could mean barriers to entry for AV ownership will decrease.

Given all of the above, when an individual consumer is presented with the option to purchase an AV, and then monetize that owned asset when not in use for little effort, wouldn't he or she act in their own self-interest and choose this? As long as he or she has a parking space for the AV in the evening or when demand is low, would this not be a more rational choice, economically speaking, than purchasing trips on-demand?

Rather than the majority of people becoming consumers of mobility and purchasing trips on-demand in an automated future, it's possible that the vast majority of consumers could be incentivized to own an AV, and also try and capture some of the value associated with automation by sharing their owned AV when not in use. Only in circumstances where land density and economics make owning/storing a vehicle prohibitively expensive would this not be the case. The future of AV ownership could therefore be much more complex and be comprised of many individual vehicle owners alongside small-scale and large-scale fleet operators.

What this means is that, if left to the free market, the convergence of EASC and MaaS may not occur in a way that benefits cities. As shown in Figure 2, rather than the cost per mile (price) being the dependent variable and the ownership model being a binary choice between individual AV ownership and fleet ownership where EASC and MaaS converge, EASC and MaaS convergence may instead be a complex outcome in constant flux and will be different in each urban market. Convergence may instead occur dynamically as a function of the mix of ownership models in any given market and people's propensity to share vehicles and rides.

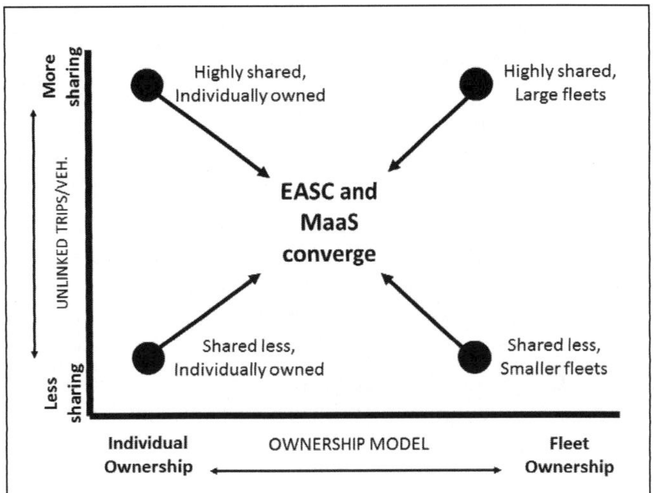

Figure 2. An alternative mobility model of the future.

Put differently, EASC and MaaS convergence may not be the recipe to achieve an end state, rather it could be the outcome of the ownership/sharing recipe in any given urban market. As such, policymakers and practitioners who are seeking to realize the potential benefits of vehicle automation for urban mobility and maximize equity, efficiency, and equity for cities need to consider governance of AV ownership and the means of vehicle sharing (i.e., ridesourcing platforms).

An immediate question is: to what extent do consumers want to share their owned AVs? The current literature includes surveys that study people's attitudes towards AVs; however, the focus of these surveys has largely been on people's perceptions of comfort, openness, familiarity, and the sociability of driving (*21*). The author was unable to find any market research gauging people's interest in owning one or more AVs and then monetizing them by sharing them when not in use.

Another way to gauge consumers' interest in owning and sharing is to consider the experience of current carsharing business models which permit this, such as Turo and Flightcar. These companies allow individuals to share their owned vehicles (albeit, non-automated) when not in use (*22*). Thus far, they have not yet achieved the level of scale of other sharing platforms, such as Airbnb, due to a variety of problems related to liability, insurance, and poor maintenance (*23*). However, many of the problems facing Turo and Flightcar have been due to human factors that will likely be removed once driving is automated. As such, the current experience of these business models may not yet be translatable to what will occur once vehicle automation reaches scale.

The only instance in the literature found by the author where there is an understanding that sharing of individually owned AVs could occur is in Elon Musk's Master Plan Part II, where he states:

When true self-driving is approved by regulators, it will mean that you will be able to summon your Tesla from pretty much anywhere. Once it picks you up, you will be able to sleep, read or do anything else enroute to your destination. You will also be able to add your car to the Tesla shared fleet just by tapping a button on the Tesla phone app and have it generate income for you while you're at work or on vacation, significantly offsetting and at times potentially exceeding the monthly loan or lease cost. This dramatically lowers the true cost of ownership to the point where almost anyone could own a Tesla. Since most cars are only in use by their owner for 5% to 10% of the day, the fundamental economic utility of a true self-driving car is likely to be several times that of a car which is not. (*24*)

Musk's concept of the future is therefore consistent with the premise in this paper: there will likely be compelling incentives for consumers to share their individually owned AVs. It should be noted though that Musk is specific in his statement about referring to the vehicles as being the "Tesla shared fleet" and the ridesourcing platform being the "Tesla phone app". Indeed, Tesla has banned its vehicles from being used on Lyft and Uber's platforms.

Musk's statement suggests that Tesla may be positioning to both supply the vehicles for EASC as well as create the future platform for MaaS. In the digital economy, the platform is the business; if Tesla eventually supplies vehicles and limits their use to its own platform, the network effects may result in Tesla becoming most dominant in this space. The outcome would be Tesla having a great deal of control and influence over how EASC and MaaS converge, all while having individual vehicle owners bear most of the burden of daily maintenance and issues that go along with owning and operating a vehicle fleet.

Most of the other leading companies at the forefront of AV deployment, such as Uber, Waymo, Lyft, Ford, GM, and Daimler have been less clear about their visions for EASC and MaaS convergence. Nearly all of them have alluded to a future of large shared fleets; however, it is unclear whether they will also allow individual AV ownership and sharing, as Tesla and Musk have proposed.

Considerations for Policymakers and Practitioners

If it is ultimately possible for consumers to simultaneously own and share an AV—and Musk has suggested that Tesla will make this possible—the incentives for consumers to simultaneously own and share could be high. This could mean that individual AV ownership could be highly disseminated and the majority of AVs could be owned by individuals and small fleet operators. As stated previously, rather than EASC and MaaS convergence being the recipe to achieve an end state, they may remain distinct paths with the ownership/sharing recipe in any given market being what ultimately influences urban mobility outcomes.

The risk facing policymakers and practitioners is the possibility of a "race to the bottom," where the average number of vehicles per household increases and multiple proprietary ridesourcing platforms, which may have their owned shared fleets as well, continually drive down and/or even subsidize the cost of rides to gain market share. This scenario would have major implications for cities and the ongoing successful and efficient function of their transportation systems. There is already a mismatch between supply and demand of road capacity and curbside space in major cities; without good governance of AV ownership and ridesourcing platforms, this mismatch could be exacerbated by highly inefficient trip-making and lower trip optimization across the entire transportation network which could result in more cars, more congestion, more parking etc. AVs could make the transportation network worse with (lack of) governance of AV ownership and ridesourcing platforms being the primary determinants.

If the benefits and potential of an automated future are to be realized, policymakers and practitioners will need to gain a better understanding of the influence and interplay of these governance issues on urban mobility outcomes.

The following are potential questions to help frame these future discussions:

1. Will any individual and/or entity be allowed to own and/or share an AV and/or fleet of AVs in future? What is the right approach for good governance of AV ownership to achieve the most desirable urban mobility outcomes?
2. Will ridesourcing platform data collection and use remain proprietary in future? What is the right approach for good governance of the means for vehicle sharing to achieve the most desirable urban mobility outcomes?

As a preliminary response to the first question, it seems unlikely and infeasible that government officials would ever be able or willing to restrict AV ownership and/or sharing. Democracy and capitalism would likely never permit such overarching control on ownership rights. This is not to say that policymakers couldn't strictly manage the ownership, use, and sharing of an individually owned AV through a strong regulatory framework; indeed, the literature already includes ideas about how partnerships could work for large commercial shared fleets, where government sets performance standards and pricing as well as provides oversight to ensure compliance. In these scenarios, municipal and state transportation agencies could become mobility authorities and the role of government becomes one of connecting control of roads, mass transit, and operators, and then charging access to those networks. Pricing mechanisms would likely need to be introduced to manage scarcity and/or oversupply and policymakers would need to ensure these pricing mechanisms are set up in a way so that they manage both the supply of vehicles as well as the supply of trips. Although they would likely be unpopular, policies and regulatory mechanisms are available that policymakers could introduce to manage the effects of individual AV ownership and sharing of them when not in use; however, future scenario-building and modeling should give greater consideration to sharing of individually owned AVs (rather than just commercial shared AV fleets) as a fundamental consideration when developing policies and strategies.

As a preliminary response to the second question, if convergence of EASC and MaaS is indeed the model to maximize access, efficiency, and equity across urban transportation networks, then the question of how to achieve good governance of ridesourcing platforms will need much greater consideration by policymakers. Ridesourcing platforms are the means by which sharing of rides and vehicles occurs, and although some of the literature alludes to greater coordination and partnerships with ridesourcing platforms when it comes to data sharing, the author was unable to find any explicit mention of the implications of ongoing ridesourcing platform data propriety for future urban mobility outcomes. Given that in a digital economy, the platform is the business, if the predominant way people source their mobility in the future is in the hands of multiple private, proprietary platforms, and these platforms continue to operate in a way where they do not communicate with one another and pick and choose what data they share with government agencies, it seems unlikely that EASC and MaaS will converge in a way that maximizes benefits for cities.

Transportation policymakers and researchers can look to the early days of the internet for a possible policy response to ridesourcing platform data propriety. Very early in the internet's development, the concept of "net neutrality" was established, a principle that ensured internet service providers (ISPs) enabled access to all content and applications regardless of the source, and without favoring or blocking particular products or websites (*25*). Policymakers might consider whether the principle of net neutrality is translatable to the future governance of ridesourcing platforms.

What this could mean is that in similar fashion to how ISPs must be neutral in enabling all content to their users, policymakers could consider if governance and function of ridesourcing platforms could somehow incorporate a similar principle of "ridesourcing platform neutrality". This would mean that similar to ISPs being neutral in enabling all content on the internet to their users, ridesourcing platforms would be neutral in enabling access to and showing all content (i.e., commercial trips available) in the network, regardless of the source, and without favoring or blocking particular product or ride.

Taking this principle one step further, it might also be possible to formalize neutral communication protocols between ridesourcing platforms to ensure that each company's algorithms for ridepooling communicate with one another and work together. This would ensure that customers using different platforms could still share a ride together. In partnership with government, these protocols could also ensure that ridesharing/sourcing data collected by the platforms are safeguarded for privacy and shared in a manner for overall optimization of each city's transportation network. From an economic perspective, these protocols might also be beneficial by encouraging value creation, because ridesourcing platforms would have to create their value through data mining rather than rent-seeking through data hoarding. In this way, competition (and value creation) would continue and be encouraged, but within a framework that safeguards access, efficiency, and equity (and privacy).

Conclusion

AV technology presents an opportunity to remake urban mobility in a way that maximizes access, efficiency, and equity. One of the roles for policymakers is to ensure that future governance of AVs achieves this.

When considering governance, the current literature centers on issues related to the safe operation and deployment of AVs but has not fully considered the implications of AV ownership and ridesourcing platform data propriety on achieving the most desirable urban mobility outcomes.

Specifically, the literature has not considered the following possibilities:

1. A future scenario where individually owned AVs are shared when not in use; and
2. The implications of ridesourcing platform data remaining proprietary in future.

This paper has found that incentives for individuals to both simultaneously own and share an AV (or indeed multiple) could be high, resulting in AV ownership being highly disseminated. The majority of AVs could be owned by individuals and small fleet operators. Disseminated ownership may not preclude large-scale sharing and the extent to which individually owned AVs are shared could ultimately determine our mobility future. EASC and MaaS are distinct and rather than their convergence being the recipe to achieve the most desirable urban mobility outcomes, their convergence may instead be an outcome of the ownership/sharing recipe in any given urban market. Policymakers and practitioners should therefore pay attention to good governance of AV ownership and the means of sharing (i.e., ridesourcing platforms).

If the future holds disseminated AV ownership and ridesourcing platform data propriety, the risk facing policymakers could be a "race to the bottom" where average number of vehicles per household keeps increasing and multiple proprietary ridesourcing platforms encourage inefficient tripmaking to gain market share. More vehicles and more trips would translate into undesirable urban mobility outcomes and it is unlikely that the possible benefits of vehicle automation would be realized.

The premise of this paper is that how policymakers govern AV ownership and ridesourcing platforms will be the major determinant in achieving the most desirable urban mobility outcomes. In order to safeguard access, efficiency, and equity, as well as harness the potential of a convergence of EASC and MaaS, additional future scenario-building and modeling will need to consider the implications of these issues of governance and what tools and policy levers would be most appropriate for good governance of AV ownership and ridesourcing platforms.

This paper has provided some early ideas for managing the effects of simultaneous individual ownership and sharing of AVs through regulatory and pricing mechanisms. It has also provided some early ideas and principles for a future regulatory framework for ridesourcing platforms that would encourage competition yet still safeguard access, efficiency, and equity by considering how the principle of net neutrality could be applied to the ridesourcing platform space.

The risk for policymakers is this: EASC and MaaS may not converge in a way that benefits cities and the questions surrounding AV ownership and ridesourcing platform data propriety are key. If the widely touted benefits of AVs, such as fewer vehicles on our roads, reduced greenhouse gas emissions, increased access for vulnerable populations, affordable housing, and a reduced need for parking are to be realized, then an increased understanding of how governance of AV ownership and ridesourcing platform data propriety could impact transportation network in cities and urban mobility outcomes should be an immediate priority for transportation researchers, policymakers, and practitioners.

Acknowledgments

The author wishes to acknowledge and thank his employer, the City of Mississauga, for supporting his interest in vehicle automation. Thank you to Michelle Berquist for the "race to the moon" analogy. Thank you also to Brendon Hemily, Bern Grush. and the Standing Committee on Strategic Management for the very helpful and constructive feedback on the initial draft. Lastly, the author wishes to thank his wife and daughter, for always being willing sounding boards on obscure ideas about an uncertain future.

Author Contributions

The author confirms sole responsibility for the following: study conception and design, data collection, analysis and interpretation of results, and manuscript preparation.

References

1. Foxx, A. Keynote Address. *Transportation Research Circular Number E-C222: Proc., Automated Vehicles Symposium*

2016. Transportation Research Board of the National Academies, Washington, D.C., 2016.

2. United Nations Educational, Scientific and Cultural Organization. Concept of Governance. http://www.unesco.org/new/en/education/themes/strengthening-education-systems/quality-framework/technical-notes/concept-of-governance/. Accessed July 2017.
3. Lee, T. Car companies' Vision of a Gradual Transition to Self-driving Cars has a Big Problem. *Vox Magazine*, 2017. https://www.vox.com/new-money/2017/7/5/15840860/tesla-waymo-audi-self-driving. Accessed July 2017.
4. Bloomberg Philanthropies, Aspen Institute. *Taming the Autonomous Vehicle: A Primer for Cities*, 2017. https://www.bbhub.io/dotorg/sites/2/2017/05/TamingtheAutonomousVehicleSpreadsPDFreleaseMay3rev2.pdf. Accessed June 2017.
5. National Association of City Transportation Officials. *Federal Automated Vehicles Policy: Accelerating the Next Revolution in Roadway Safety*, 2016. https://www.transportation.gov/sites/dot.gov/files/docs/AV%20policy%20guidance%20PDF.pdf. Accessed June 2017.
6. MAASiFiE. About MaaS Concept and MAASiFie Project, 2016. http://www.vtt.fi/sites/maasifie. Accessed October 2017
7. Chase, R. What Uber's Next CEO Needs to Say. July 5, 2017. *Issue of Wired*. https://www.wired.com/story/what-ubers-next-ceo-needs-to-say/. Accessed July 2017.
8. Qinglu, M., S. Kockelman, and M. Segal. *Making the Most of Curb Spaces in a World of Shared Autonomous Vehicles: A Case Study of Austin Texas*, 2017. http://www.caee.utexas.edu/prof/kockelman/public_html/TRB17ReusingCurbParking.pdf Accessed July 2017.
9. Brookings Institution. *Moving Forward: Self-Driving Vehicles in China, Europe, Japan, Korea and the United States* (West, D. M., ed.), 2016. https://www.brookings.edu/research/moving-forward-self-driving-vehicles-in-china-europe-japan-korea-and-the-united-states/. Accessed June 2017.
10. Bloomberg New Energy Finance, and McKinsey & Company. *An Integrated Perspective on the Future of Mobility*, 2016. https://data.bloomberglp.com/bnef/sites/14/2016/10/BNEF_McKinsey_The-Future-of-Mobility_11-10-16.pdf. Accessed June 2017.
11. Boston Consulting Group. *Revolution in the Driver's Seat: The Road to Autonomous Vehicles*, 2015. https://www.bcg.com/publications/2015/automotive-consumer-insight-revolution-drivers-seat-road-autonomous-vehicles.aspx. Accessed June 2017.
12. Cohen, T., and C. Cavoli. Automation of the Driving Task: Some Possible Consequences and Governance Challenges. *Prepared for the Roundtable on Cooperative Mobility Systems and Automated Driving*, 6–7 December 2016. https://www.itf-oecd.org/sites/default/files/docs/automation-driving-consequences-governance-challenges.pdf. Accessed June 2017.
13. The European Parliament, Directorate-General for Internal Policies of the Union. *Self-Piloted Cars: The Future of Road Transport?* Research for TRAN Committee, 2016. http://www.europarl.europa.eu/RegData/etudes/STUD/2016/573434/IPOL_STU(2016)573434_EN.pdf. Accessed June 2017.
14. Forbes. *10 Million Self-Driving Cars Will Hit the Road by 2020 – Here's How to Profit* (Garret, O., ed.), 2017. https://www.forbes.com/sites/oliviergarret/2017/03/03/10-million-self-driving-cars-will-hit-the-road-by-2020-heres-how-to-profit/#75d4e1917e50. Accessed June 2017.
15. Morgan, S. *Shared Mobility on the Road of the Future*, 2016. http://www.morganstanley.com/ideas/car-of-future-is-autonomous-electric-shared-mobility. Accessed Jul 2017.
16. Anderson, J. M., N. Kalra, K.D. Stanley, P. Sorensen, C. Samaras, and T.A. Oluwatola. *Autonomous Vehicle Technology: A Guide for Policymakers*. RAND Corporation, RR-443-2-RC, Santa Monica, CA, 2016.
17. Rocky Mountain Institute. *Peak Car Ownership: The Market Opportunity of Electric Automated Mobility Services* (Johnson, C., and Jo. Walker, eds.), 2016. http://www.caee.utexas.edu/prof/kockelman/public_html/TRB17ReusingCurbParking.pdf. Accessed June 2017.
18. UITP/International Association of Public Transport. *Autonomous Vehicles: A Potential Game Changer for Urban Mobility*, 2017. http://www.uitp.org/sites/default/files/cck-focus-papers-files/PolicyBrief_Autonomous_Vehicles_LQ_20160116.pdf. Accessed July 2017.
19. United States Department of Transportation. *Beyond Traffic 2045*, 2016. https://www.transportation.gov/sites/dot.gov/files/docs/BeyondTraffic_tagged_508_final.pdf. Accessed July 2017.
20. Arbid, J., and T. Seba. *Rethinking Transportation 2020-2030. The Disruption of Transportation and the Collapse of the Internal-Combustion Vehicle and Oil Industries*, 2017. https://static1.squarespace.com/static/585c3439be65942f022bbf9b/t/591a2e4be6f2e1c13df930c5/1494888038959/RethinkX + Report_051517.pdf. Accessed June 2017.
21. London School of Economics. Autonomous Vehicles: Negotiating a Place on the Road (Tennant, C., et al., eds.), 2016. www.lse.ac.uk/website-archive/newsAndMedia/PDF/AVs-negotiating-a-place-on-the-road-1110.pdf. Accessed July 2017.
22. Walker, J. The Traffic Jam of Robots: Implications of Autonomous Vehicles on Trip-Making. *Transportation Research Circular Number E-C22: Proc., Automated Vehicles Symposium 2016*. Transportation Research Board of the National Academies, Washington, D.C., 2017.
23. Yahoo Travel. *Car-Sharing Companies Make You Money While You're Parked at the Airport*, 2014. https://www.yahoo.com/style/car-sharing-companies-make-you-money-while-youre-88410906812.html. Accessed October 2017.
24. Musk, E. *Master Plan, Part Deux*, 2016. https://www.tesla.com/en_CA/blog/master-plan-part-deux. Accessed July 2017.
25. Oxford Dictionaries. https://en.oxforddictionaries.com/definition/net_neutrality. Accessed July 2017.

The Standing Committee on Strategic Management (ABC10) peer-reviewed this paper (18-03403).

Article

TRR
JOURNAL OF THE TRANSPORTATION RESEARCH BOARD

Transportation Research Record
2018, Vol. 2672(7) 24–33
© National Academy of Sciences:
Transportation Research Board 2018
Article reuse guidelines:
sagepub.com/journals-permissions
DOI: 10.1177/0361198118796968
journals.sagepub.com/home/trr
SAGE

Autonomous Transport: Transforming Logistics through Driverless Intelligent Transportation

Anders Hjalmarsson-Jordanius[1], Mikael Edvardsson[2], Martin Romell[2], Johan Isacson[3], Carl-Johan Aldén[4], and Niklas Sundin[5]

Abstract

How can autonomous technology be used beyond end-customer autonomous driving features? This position paper addresses this problem by exploring a novel autonomous transport solution applied in the automotive logistics domain. We propose that factory-complete cars can be transformed to become their own autonomous guided vehicles and thus transport themselves when being moved from the factory for shipment. Cars equipped with such a system are driverless and use an onboard autonomous transport solution combined with the advanced driver assistance systems pre-installed in the car for end-customer use. The solution uses factory-equipped sensors as well as the connectivity infrastructure installed in the car. This means that the solution does not require any extra components to enable the car to transport itself autonomously to complete a transport mission in the logistics chain. The solution also includes an intelligent off-board traffic control system that defines the transport mission and manages the interaction between vehicles during systems operation. A prototype of the system has been developed which was tested successfully in live trials at the Volvo Car Group plant in Gothenburg Sweden in 2017. In the paper, autonomous transport is positioned in between autonomous guided vehicles and autonomous driving technology. A review of the literature on autonomous vehicle technology offers contextual background to this positioning. The paper also presents the solution and displays lessons learned from the live trials. Finally, other use areas are introduced for driverless autonomous transport beyond the automotive logistics domain that is the focus of this paper.

Car manufacturers (original equipment manufacturers (OEMs)) will in the near future produce cars with autonomous vehicle technology (AVT), enabling advanced autonomous driving (AD) capabilities. The proposed benefits of AVT are numerous, including increased road safety as well as the convenience and freedom of users to choose what to do during their travel time instead of driving the vehicle (*1*, *2*). A more efficient traffic system and assumed mitigation of environmental impact through optimized fuel consumption are other anticipated benefits (*3*). Yet AVT may possess more value in other areas beyond end-customer applications and benefits to society in general (*4*).

This paper addresses how AVT may create benefits in the early phases of the automotive logistics chain. It discusses how AVT can be used to transform cars to become their own means of transportation and transport themselves, driverless, from the point where they leave the factory to the loading location for long-haul transportation. This system would transform the car to a driverless *autonomous transport* (AT) vehicle moving itself without a human driver during the first logistical steps after being released from a production plant. The purpose of this position paper is to introduce AT as a concept and illustrate its characteristics with the prototype developed in Born to Drive. The paper also aims to identify and explore the strengths, weaknesses, threats, and opportunities of AT in the mobility domain: *What constitutes an autonomous transport solution implemented in a car?* And *What are the strengths, weaknesses, opportunities, and threats to an autonomous driverless transport solution?*

[1]Digital Innovation at RISE Viktoria and University of Borås, Gothenburg, Sweden
[2]Volvo Car Group, Gothenburg, Sweden
[3]Combitech AB, Gothenburg, Sweden
[4]Semcon Sweden AB, Gothenburg, Sweden
[5]Consat Engineering AB, Gothenburg, Sweden

Corresponding Author:
Address correspondence to Anders Hjalmarsson-Jordanius: Anders.hjalmarsson@ri.se

The paper is organized as follows. In the second section, AT is defined in comparison with autonomous driving and autonomous guided vehicles. In the third section, Born to Drive is described as a case study, with a detailed account of the project setup, the research design, and the solution domain. In the fourth section, the AT solution is presented, highlighting design features and system components. The fifth section presents lessons learned from the full-size live car trials performed in 2017. The paper concludes with a discussion in which key aspects of the solution potential beyond automotive logistics are elaborated upon, as well as a summary of conclusions, and topics for future research.

Current Applications and Key Concepts

In this section, current applications of autonomous technology for logistics are explored. This is followed by a positioning of the AT concept in relation to AD and autonomous guided vehicles (AGV).

Current Applications of Autonomous Technology in Logistics

Heutger and Kückelhaus (*5*) provide an in-depth analysis of the implications AVT may have for logistics. They argue that there are strong arguments for suggesting that the logistics industry will adopt AVT faster than other industries. The reason for this is that different rules may apply when a vehicle is driving in a secure, cloistered zone, compared with vehicles that operate in the open traffic system. In addition, liability issues are less pressing when goods are moved by AT solutions instead of people. These reasons support the prediction that organizations involved in the logistics chain will adopt AVT faster than, for example, public transportation on a large scale (*5*).

In the analysis put forward by Heutger and Kückelhaus (*5*), they argue that the numerous examples of AVT already in operation provide evidence that driverless vehicles are safe and successful in closed environments, and that they have the potential to move into other areas within the logistics industry not yet targeted with AVT. For example, self-driving vehicles today often move materials in private warehouses and controlled open-air sites. Beyond warehousing operations, Heutger and Kückelhaus (*5*) expect many more applications in future along the entire logistics chain, particularly in outdoor logistics operations, line-haul transportation, and last-mile delivery.

Despite the comprehensive analysis of different examples of AVT being used to change the logistics industry by Heutger and Kückelhaus (*5*), the concept of the freight-being-moved transporting itself is not covered.

Key Concepts: Autonomous Guided Vehicles, Autonomous Drive, and Autonomous Transport

Using AVT to enable freight-being-moved to transport itself is relatively similar in principle to the technology used in AGV. Three examples to illustrate the latter technology initiate this section.

The first example to illustrate AGV technology is the solution developed for parking and arranging vehicles at airports (*6*) (Example A). The AGV in this example picks up the car and transports it to the correct position in the parking garage. This means that any car, not just a car with a high technological level, can be transported without an operator. This solution may be applicable for transporting niched vehicles within car manufacturing as well, for example, between two assembly lines (*6*). This AGV however requires that there are no mixed or crossing tracks (*6*) reducing flexibility and limiting the applicability of the system.

A second example of using AGV is the use of autonomous orchard vehicles adopted by fruit production companies (*7*) (Example B). Here a car is rebuilt and equipped with systems giving it the ability to drive by itself so that the vehicle becomes an AGV. The vehicle is then able to navigate within a limited area without a driver to fulfill a specific mission. The drawback here is that the car needs to be adapted in terms of infrastructure to operate.

The third example of AGV is that applied by Nissan in which autonomous tow trollies carry production-complete cars between designated loading and unloading points at manufacturing plants using a modified Nissan LEAF as head vehicle (*8*) (Example C). Unlike conventional AGV for transporting parts, which often require the installation of rails or extensive use of magnetic tape, this system does not need any special infrastructure and instead relies completely on the AD capabilities installed in the towing car. The towing car is equipped with an array of additional cameras and laser scanners that detect lane markings, curbs, and potential obstacles or hazards around the vehicle (*8*).

In summary, an AGV is a driverless vehicle often used for internal logistics, including material handling, and is mostly used within manufacturing industries. The AGV drives automatically and can in a controlled fashion be used in environments where humans also work. Over the years many types of AGV have been developed of different size or speed, and various navigation systems controlling how they interact with their surrounding environment (*5*).

Compared with AGV, AD can be viewed as an end-customer application of AVT that allows vehicles (e.g., cars or trucks) to make their own decisions based on inputs from the surrounding environment. AD vehicles are graded on a scale of five formal levels of automation;

the three highest levels indicate that the vehicle operates without human intervention, but with a human present in the vehicle as driver or passenger.

The term autonomous transport (AT) was coined during the Born to Drive project to denote AVT utilized for use areas other than end-customer AD solutions (e.g., driver support). AT as a concept is positioned in this paper as a combination of AD and AGV technology. An AT equipped car with this mode turned on is relatively similar to an AD car, with the exception that it is being moved driverless, with no human present in the car to complete a logistical mission. The mission is a transport order to move the vehicle not as a machine to move people or goods, but instead move itself as freight-being-moved from point A to point B. As no human is involved either as driver or present in the vehicle, the vehicle has to be linked to an intelligent traffic control system that provides the transport mission and coordinates the vehicle in relation to other vehicles throughout the transport route. An AT equipped car with the solution enabled thus has strong similarities with an AGV as it completes a logistical mission with AVT. Consequently, an AT-enabled car can be viewed as an AD car acting as its own AGV during transportation in a logistics situation (*4*). AT in this paper is thus defined as *a vehicle, acting as its own AGV, transporting itself driverless using AD software and intelligent transport technology, to execute, coordinate, and complete a logistical mission.*

Figure 1 positions AT in the crossover between AD and AGV. Examples A, B, and C are included in the figure to map these solutions with AD, AGV, and AT. Example D in Figure 1 represents the Born to Drive solution presented in this paper to illustrate AT in a production-complete car.

Case Presentation

This section presents the two-year R&D project Born to Drive. It also provides an account of the research design applied during the project. It is concluded by a description of the domain for which the AT solution was developed. This domain should not be viewed as a unique application area for the solution. Other application areas are discussed in the final sections of the paper.

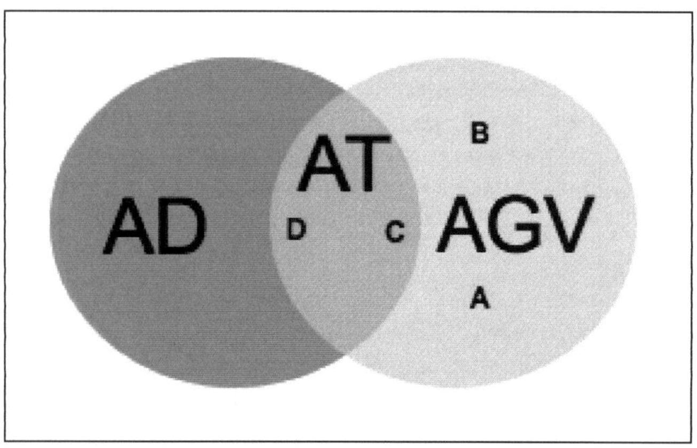

Figure 1. Position of AT in the crossover between AD and AGV.

Project Setup

Born to Drive was a two-year R&D project with the overall vision to "change the future of logistics." The main objective was to demonstrate the possible applications of AT of cars from the manufacturing plant to the outbound logistics parking yard, within a controlled area, in preparation for further long-haul transportation. The long-term objective was to develop and validate a logistics system that enables fully autonomous displacement of cars from the factory to be loaded onboard RORO ships, cargo trains, or carrier trucks. The main constraint for the novel digital innovation was that minimal additional technology (e.g., sensors, infrastructure) can be added to the car for the solution to function. Instead the solution should be based on the technology that the production car has installed when it is released from the production facility. The immediate beneficiary of the Born to Drive prototype was consequently outbound logistics (OBL). OBL is an organizational unit in the automotive logistics chain that manages the processes taking place after a produced car is cleared for delivery from the production plant. Current challenges in the OBL domain, for example, are congestion and safety. Such areas could be improved by introducing an AT solution to perform various types of movements of the car in the yard, such as maneuvering, repositioning, and loading/unloading of the vehicle.

The Born to Drive project (2015–2018) was organized as a joint R&D effort with eight work packages with different purposes, ranging from pre-development analysis, to onboard and off-board systems development, to demonstration preparation and assessment by appropriate regulatory bodies. The project involved an open innovation alliance with eight engineering companies and research institutes connected to the automotive cluster in Sweden.

Research Design

The project relied on the action design research (ADR) methodology (*9*) to structure the logic of the project as a case-based R&D study (*10*). There are several arguments that motivate the selection of ADR as the appropriate methodology. ADR provides excellent support for how to: (a) orchestrate collaboration between design researchers and business partners; (b) organize processes of knowledge co-production; (c) pursue development and

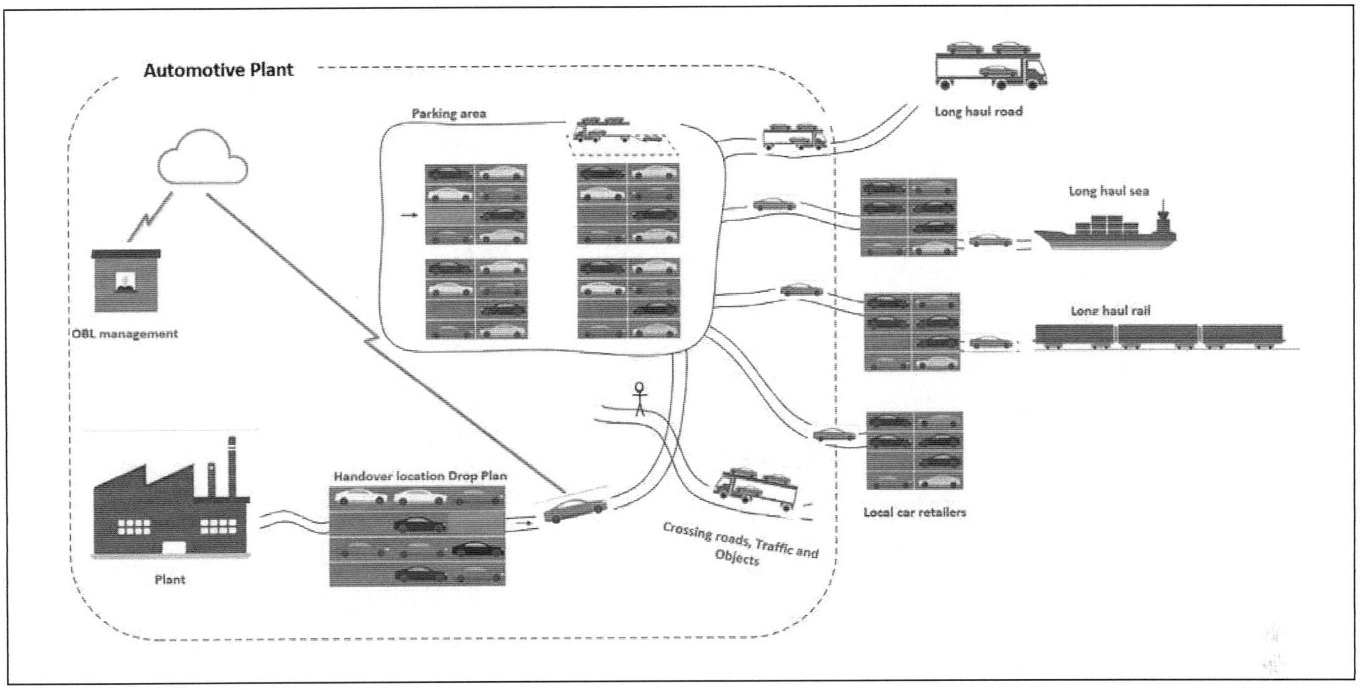

Figure 2. Solution domain: outbound automotive logistics.

evaluation activities, and (d) analyze and diffuse theoretical and practical outcomes (*8*). More specifically, it stipulates a set of guidelines, formulated as principles, to assist researchers and practitioners in their efforts to balance practical problem-solving and scientific knowledge production. Sein et al. (*9*) present an exhaustive description of the principles. These principles have served as a baseline for the project plan as well as the co-production between the business partners and the researchers. The ADR principles were intertwined with novel management principles and a structured agile engineering approach to build and validate the AT solution.

The solution was integrated in a factory-complete car, a Volvo XC90 T8 with shift-by-wire and advanced driver assistance systems (ADAS) end-customer installation. Data collection and analysis in the research project was performed using qualitative research techniques (e.g., interviews, document studies, observations). However, to measure and verify the capacity of the Born to Drive solution implemented in the car, quantitative data gathering was used and statistical analysis performed.

In this position paper, qualitative data has been used from focus groups arranged as an activity after each trial to collect and assess insights (benefits, drawbacks, opportunities, and challenges) jointly identified by the participants attending the trials of the system prototype. The focus groups were organized using the guidelines provided in Krueger (*11*) and involved engineers, managers, and personnel from the organizations participating in the project, beneficiaries, and the project sponsor. Feedback was also provided through an open-ended questionnaire that participants were invited to complete.

Solution Domain

There are approximately 220 automotive plants in the world operated by different OEMs; 140 of these are located in Europe. In total, 72 million cars were produced in 2016.

The case site for Born to Drive was the Volvo Car Group production plant in Gothenburg which has a capacity of approximately 300,000 vehicles per year. Representatives from OBL at this site acted as the main requirement stakeholders for the prototype design. The cars produced are destined for both domestic and international dealers and customers. When a car has reached completion in the factory, it is said to be factory complete (FC). The FC-car is driven from the factory building, and off the production line, out to the OBL-managed yard, and then to a nearby handover location called the drop plan (see Figure 2).

This initial sequence of moving the FC-car is currently carried out by OBL personnel. Hence, several handover locations for FC-cars exist depending on their destination on the OBL yard. From these locations, the FC-cars are driven by OBL personnel to different parking areas on the yard, based on dealer destination. After the FC-cars are parked in different parking areas in the yard, they remain there for different amounts of time, depending on the carriers' lead times, financial circumstances, or diverse agreements with the OEM's global sales companies. When the FC-cars are cleared for delivery, they are picked up on trailers which drive them directly to dealers, to a harbour where they are shipped abroad via RORO ships, or to railway sidings where they are loaded onto trains and taken toward their destinations.

These initial movements of the FC-car represent only a few of the total movements for each car along the automotive logistics chain. In total, each car is moved approximately 25 times from the production line to the dealer before reaching the end-customer. Each such movement currently requires human involvement.

Born to Drive: An Autonomous Transport Solution

By introducing an AT solution in the OBL domain the vision is to create a more efficient automotive logistics chain. The need to employ drivers to move FC-cars on the outbound yard is expected to be reduced if the cars instead are moved driverless with the AT mode turned on. This will increase safety on the yard. The yard as a storage facility can in addition hold even more cars, as cars can be parked closer to each other as there is no need to leave space for a driver to open the door to get in or out for transporting or repositioning the car. This enables the factory to increase production capacity. The incidence of damage can also be reduced and the need for trailers to move the production cars for loading on to ships, trains, or long-haul trailers can be reduced. The reduced need for drivers may also reduce the OEM's labor costs, as well as initiate a shift in the role description of OBL personnel from driving assignments to quality control and operation of the AT solution.

To achieve these aims the Born to Drive team conceptualized and assessed different types of potential solution designs throughout spring of 2016. One of three potential designs was chosen. The selected design consisted of a kernel efficient software component to be implemented in the car's electronic platform. The component should be able to be switched on and off, and when switched on it will put the vehicle in AT mode. In AT mode, the car becomes its own AGV to move itself as freight to meet a transport mission: moving the car as freight from point A to point B.

The driverless movements are coordinated by an off-board traffic control system with real-time kinematic (RTK) GPS mapped sections of the yard as base. The solution uses low-speed movements to move the FC-car from the production plant to handover locations and parking areas within the automotive plant. The core digital innovation is the unique combination of existing onboard ADAS sensors with mobile technology connectivity that enables the car to communicate with the developed intelligent traffic control system. Together this new software functionality enables up to level-4 driverless movements of the FC-car when the AT mode is enabled. The solution operates without any additional sensors or functionality than what is originally equipped in the FC-car. The solution consequently uses the available sensors in the FC-car, fusion of sensor data captured by the car, and connectivity utilizing existing and future mobile networks communication technology.

An intelligent off-board traffic control system manages each driverless FC-car now being transported as freight in the logistics flow and directs the car from A to B, as well as managing the interaction between cars in the flow. By cross-referencing data collected from the FC-cars' sensors with transport mission data, the FC-car with the AT solution enabled calculates its location, negotiating the route to its destination unaided by human agency, travels within the speed limits of the OBL yard, and automatically stops if it detects an object or any hazard ahead. The utilization of the traffic control system allows flexibility, for example, to accommodate changes in production processes, re-arrangement of the routes on the OBL yard, or changes in the structure of the parking lot. The traffic control system monitors the location of the FC-cars with the AT solution enabled, as well as any FC-cars without the AT solution but equipped connectivity technology. The system monitors speed and other information, such as remaining fuel, electric power level, logistics data, and operational status of each car. As the yard is divided into sections, this enables the system operator to close down parts of the yard and automatically re-route the vehicles in the logistics flow while executing individual transport missions. In summary, once the AT solution is turned on it will remove the need for a driver to (a) transport the car from the factory to the handover location drop plan and (b) move the car from the drop plan to other locations within the yard. In a possible future scenario, the AT solution may also move cars driverless from the OBL parking yard to loading on trains or ships for long-haul transport.

The following step-by-step sequence describe the AT solution in operation on a systems level (see Figure 3). A detailed account of the solution from a technical viewpoint is provided in Hjalmarsson et al. (*12*).

A digital transport mission with information on where the car should be parked in the OBL yard is generated from an external logistics system (step 1). This order acts as a trigger to the off-board traffic control system to automatically design an optimal transport route for the car using off-board intelligence; this route is communicated to the car (step 2), which triggers the car to transform itself to its own AGV and transport itself driverless from the production plant to the OBL parking yard (step 3). The car is positioned using satellite position technique

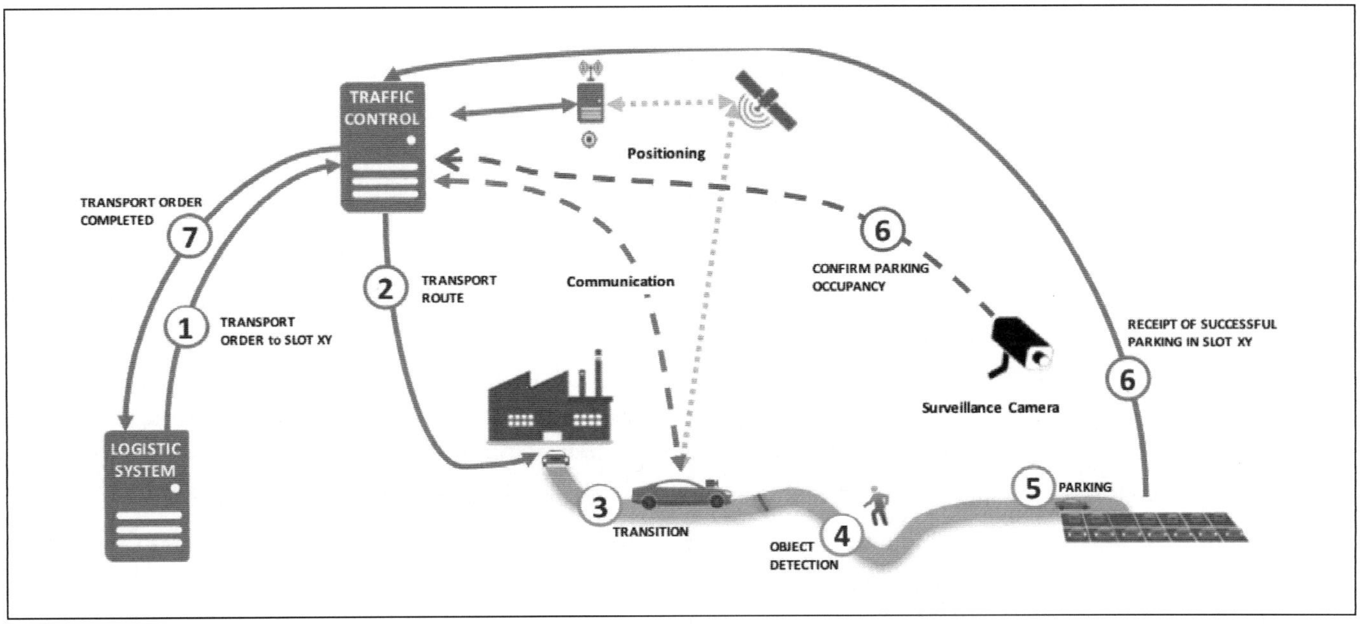

Figure 3. Overview of the AT solution.

and communicates with the traffic control system using existing and future mobile communications technology. Throughout this step, the car is connected with the traffic management system via a MQ Telemetry Transport (MQTT) communication link. In short, the MQTT broker manages the message transmission between cars and the intelligent off-board traffic control system enabling coordination and assignment of routes and instructions (*12*). The solution onboard includes functionality using the car's own sensors to detect objects on the transition routes, and functionality for making rapid decisions on how to react if objects emerge in its path (step 4). Surveillance cameras may be used as a complement to enable human agency in the otherwise fully digital solution. OBL yard handlers may consequently monitor the system in operation and intervene if necessary. The car performs parking autonomously using the AT solution complemented by the onboard ADAS functionality installed for the end-customer (step 5). The system sends a receipt to the traffic management system when the parking is completed (step 6), which in turn automatically reports to the logistics system that the transport order has been performed (step 7).

Live Solution Demonstrations: Design and Lessons Learned

During spring and fall of 2017, the implemented AT-solution prototype was demonstrated in three full-size car demonstration events. The first two-hour event was performed in May 2017 and involved the project team. The second and third demonstrations took place at the Volvo Car Group manufacturing plant in Sweden, in the operational environment for an industrialized solution aimed at an OBL application. The events attracted in total 90 participants from the involved project partners, as well as representatives from beneficiaries, the project sponsors, and invited stakeholders. The aim of the live car trials was: to prove the concept in an operational setting, to demonstrate to senior management the AT solution as a key component in a complete vision of how to use AVT in automotive logistics to achieve Industry 4.0 impact, and to learn about the benefits, challenges, and opportunities of the AT solution.

Demonstration Design

For the demonstration events, a traffic control room was constructed to illustrate a future possible work environment for the yard handlers operating the system. This room included a user interface for the traffic control system, a connection to the logistics system, and data generated by the sensors in the AT solution onboard the trial car. The trial began in this room with a presentation of the project, the onboard solution, and the off-board traffic control system. The participants were then invited to move to the OBL yard to experience the AT solution in action.

The demonstration in the yard covered the complete step-by-step sequence as described above. The trial began with the scenario that a transport order was distributed from the legacy logistics system to the traffic control system, and the distribution of the transport route from the

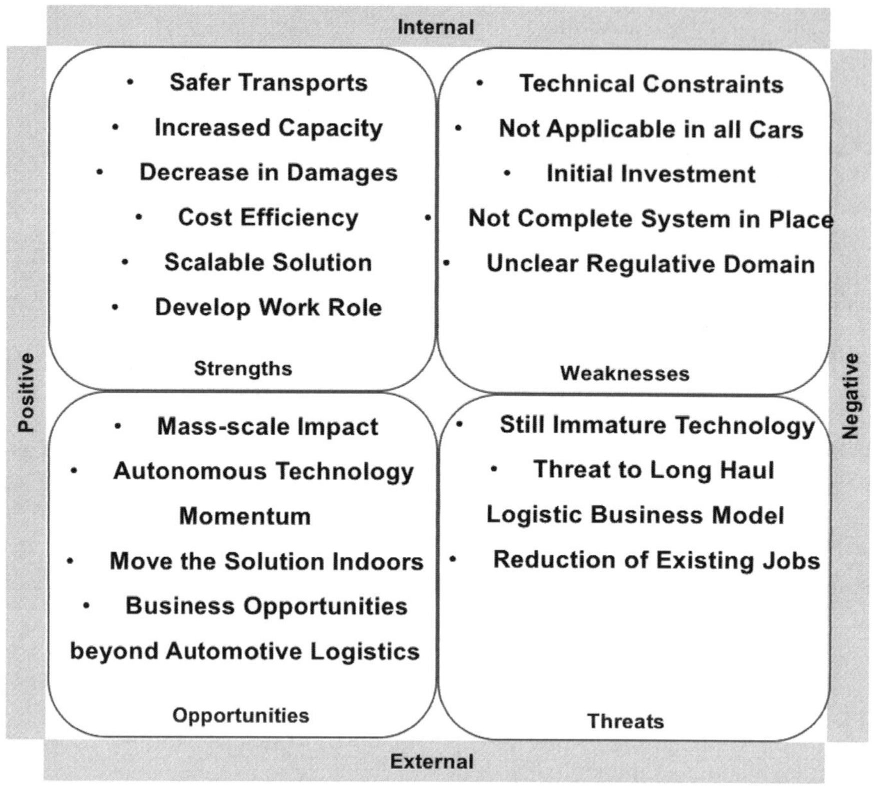

Figure 4. SWOT analysis.

traffic control system to the AT-solution enabled car. The trial was then continued with the car acting as its own AGV moving in a route from the production plant drop area to a designated parking lot on the yard and a sub-sequence in which the car parked itself in the parking lot. The trial then included a visualization of autonomous vehicle control through which the car reversed from the parking lot and continued on a second route from the first parking lot to a second parking lot. During this second transport mission, which also was initiated through a traffic control order, the onboard functionality in the system performed autonomous object detection and autonomous stop of the vehicle, using a crash-test dummy to demonstrate an object intruding in the route that the car was taking in the process of fulfilling the transport mission. After removal of the object detected, the route was completed by the car parking on the second parking lot, and sending a receipt of the successful completion of the transport mission to the traffic control system. This message was forwarded to the legacy logistics system which in turn triggered the car to switch off the AT-transport mode and thus changed the car from being its own AGV to resume the mode of being a car again.

Strengths, Weaknesses, Opportunities, and Threats of Autonomous Transport

A focus group was conducted as a final activity in each trial event to identify, discuss, and collect ideas and feedback from the participants based on their immediate experiences from the trial. The focus group was organized as a joint dialogue facilitated by one of the senior directors from the project team. The purpose was to elicit the benefits, challenges, and opportunities of the AT solution, as well as looking beyond automotive logistics to see the possibilities and barriers of the solution in other application areas. The focus groups involved all 90 participants and took about one hour each. The focus groups were followed by an open-ended questionnaire in which participants could feedback additional comments about the solution and the trial event. This generated 30 feedback questionnaires. The outcome from the focus groups was transcribed and together with the feedback questionnaires used to develop the SWOT analysis presented in Figure 4.

Strengths of the Autonomous Transport Solution. The most significant benefit of AT is the potential for a safer OBL yard when the solution is applied in the automotive logistics domain. AT-enabled FC-cars can eliminate errors made by human agency and thus increase safety. In addition, the AT solution increases the ability to control the flow of traffic. Increased capacity—owing to fitting more cars in the parking yard and reduction in time when cars are parked—are additional strengths of the AT solution. As a car is moved approximately 25 times between the production plant and the dealer this creates risks of damage in the process of moving the car, for example, when doors are opened, hitting other parked cars on the yard or objects in narrow spaces like in a train carriage. By removing the driver, the number of incidents of damage to cars between the plant and the dealer is expected to be reduced, which in turn reduces costs and the time customers wait in order to receive cars they have purchased.

The removal of the driver in the logistics chain also creates the benefit of reduced personnel costs, and also the opportunity to transform the role of driver to that of system operator and to transfer more resources to quality control. Another strength identified in the group discussion was the ease to scale the AT solution as it does not require any additional sensors or other infrastructure beyond what is already equipped in the car when it is factory complete. In a plant that produces 300,000 to 500,000 vehicles a year, additional infrastructure on each car to make an AT solution work would likely create a high add-on cost making the solution not economically justifiable. This is not the case with the Born to Drive AT solution.

Weaknesses of the Autonomous Transport Solution. The weaknesses discussed in the focus groups regarding the AT solution came in various forms. One weakness addressed was that the AT solution required ADAS functionality as well as gear shifting by wire, two features which are not yet common standard amongst the majority of cars being produced today. However, the projection is that this share will increase rapidly as AVT evolve and new car platforms are introduced. It is envisioned that by mid 2020 the majority of cars being built will fulfill the technological requirement for the AT solution developed in Born to Drive.

Another weakness is that the solution is at a prototype level and not in place as a completed solution. The AT solution has to be transformed from a prototype to a validated system solution and that requires additional development as well as further validation. However, the project has proved that existing technology can be used as a base to develop an onboard AT system with a guiding off-board traffic control system linked to an existing legacy logistics system (*12*). This proof-of-concept is a capable point of departure for completing the system to industrial implementation potential.

A third weakness was identified in relation to current regulation. As AT is defined, and positioned in the crossover between AD and AGV, it is uncertain which regulatory body should govern the vehicle when it is in AT mode—transforming the car to an AGV to execute a transportation mission. An AGV is a machine and is regulated as such. Should a car in AT-solution mode without a driver involved be viewed from the same legal stance? Or should it be viewed as an AD car and be regulated through that regulatory framework? The regulatory issue has to be resolved to make the Born to Drive AT solution ready for operation.

Opportunities of the Autonomous Transport Solution. The opportunities highlighted during the focus groups related to both the short-term application area OBL within the automotive industry and long-term opportunities outside this specific area. Taking into account that it will take some time to manage the technical constraints in current cars (e.g., lack of ADAS and shift-by-wire) the massive number of cars produced each year (72 million cars in 2016) creates an interesting opportunity to mass-scale an AT solution for OBL in the automotive industry. All cars produced need to be moved between the production plant and the dealer. Consequently a massive market is growing for AT solutions as cars become more autonomous and connected.

Moving beyond the automotive industry, the focus groups also discussed other potential markets to apply driverless AT solutions. One such business area is parking, where a driverless AT mode makes it possible to park the car autonomously in a parking garage that is governed by an off-board traffic control system. This will enable operators of parking facilities to increase their capacity by utilizing the space more efficiently. Another business area may be the car rental business which, by using the AT mode in between rentals, could make operations in the process to prepare, deliver, reposition, and store cars more efficient without the use of human agency as drivers.

One enabling condition for the AT solution as developed in Born to Drive is that it builds on existing AVT. Immense resources are currently being used to evolve autonomous technology. This means that AT solutions are being developed in a strong tailwind exploring how end-customer developed AD technology can be used in other domains.

Threats of the Autonomous Transport Solution. While the opportunities of AT solutions within the automotive industry domain and beyond are promising, threats also need be identified, minded, and mitigated. Perhaps the greatest threat is the immature technology that AVT represents. The potential for injury or damage to assets via unforeseen scenarios may exist when using AT solutions.

There is a lack of evidence that proves the projected benefits. There may be unforeseen effects on the business operators that today manage short-haul and long-haul operations in the automotive industry when driverless FC-cars transport themselves and carriers are not needed. Other threats are the lack of knowledge about the effect of the solution in terms of jobs when drivers are not needed to transport vehicles along the logistics chain. One way to counter this is to transform the current jobs of drivers to quality inspectors and systems operators.

Discussion

Born to Drive has proved that AVT can be used in novel domains beyond end-customer applications. This is a major contribution to application of AVT beyond the end-customer and to meet the societal benefits envisioned (*1*, *2*, *13*). Expanding AT beyond end-customer solutions is one of the main challenges that OEMs are facing to increase the value of evolving AVT.

As seen in the SWOT analysis, driverless AT can be an enabler of cost reductions for handling FC-cars in automotive logistics, increase safety in the OBL yard, and act as an enabler to create a business case for transforming automotive logistics on a mass scale. However, during other stages in the automotive logistics process, for example, when transportation occurs in a relatively restricted, indoor, and narrow environment—such as inside a RORO-ship—key parts of the current AVT may not be applicable. The functions of an AD-equipped car developed for road traffic and urban environments may not be directly applicable inside a building or a large RORO-ship because of the divergences between these environments. Thus, any AT solution must be adapted to the environment to enable AT in indoor facilities: the autonomous function needs to be improved to be used at very low speeds, without a driver present, but with both people and other objects close to the car as it maneuvers itself. Also, indoor positioning needs be researched and adapted so that indoor AT is possible.

A system like the Born to Drive solution, scalable to low cost with no additional infrastructure or sensors beyond what a FC-car is equipped with, creates the embryo for transforming the physical systems in today's logistics operation to Industry 4.0 standard (*14*). In addition to this, other areas beyond logistics may be seen as potential export domains for this type of AT solution. Some of these potential business areas are already identified (*2*, *13*), others need to be identified and properly understood.

Conclusions and Future Research

In this paper, a novel driverless intelligent AT solution has been presented to illustrate freight-being-moved transporting itself. The AT solution enables FC-cars to temporarily shift mode and become AGV and thus their own means of transport to complete transport missions without a human driver. This driverless AT solution will enable OEMs to reshape and optimize their logistics chains from production unit to end-customer delivery. Through undertaking an ADR approach intertwined with novel management principles including an agile engineering process, a system prototype has been designed and implemented in a Volvo XC90 FC-car. The system prototype includes a traffic control system and requires no added infrastructure to the cars. Three live car demonstration events have provided important lessons as well as a validation of the value that the AT solution may create in the automotive logistics chain and especially OBL within that chain. The various strengths and opportunities are displayed in the SWOT analysis and discussed in this paper. These experiences have also identified challenges for the AT solution, primarily connected to technological constraints, unresolved system barriers, regulatory uncertainties, and lack of understanding of how driverless AT may affect existing processes and organizations. These are examples of topics for future research with AT as the point of departure.

Acknowledgments

The authors thank all the project participants for making Born to Drive a successful project, and for their direct or indirect contributions to the paper. Participating organizations are Volvo Car Group, Actia AB, Combitech AB, Consat Engineering AB, Semcon Sweden AB, RISE Viktoria, the Swedish Transport Administration (Trafikverket), and the Swedish National Road and Transport Research Institute (VTI).

Author Contributions

The authors confirm contribution to the paper as follows: study conception and design: Anders Hjalmarsson-Jordanius, Mikael Edvardsson; data collection: Anders Hjalmarsson-Jordanius; analysis and interpretation of results: Anders Hjalmarsson-Jordanius, Martin Romell, Niklas Sundin, Carl-Johan Aldén; draft manuscript preparation: Anders Hjalmarsson, Johan Isacson. All authors reviewed the results and approved the final version of the manuscript.

References

1. Fagnant, D. J., and K. Kockelman. Preparing a Nation for Autonomous Vehicles: Opportunities, Barriers and Policy Recommendations. *Transportation Research Part A: Policy and Practice*, Vol. 77, 2015, pp. 167–181.
2. Jamthe, S. *2030 The Driverless World. Business Transformation from Autonomous Vehicles*. Amazon Digital Services LLC, Seattle, WA, 2017.
3. Broggi, A., P. Cerri, S. Debattisti, M. C. Laghi, P. Medici, D. Molinari, M. Panciroli, and A. Prioletti. PROUD-

Public Road Urban Driverless-Car Test. *IEEE Transactions on Intelligent Transportation Systems*, Vol. 16, No. 6, 2015, pp. 3508–3519.

4. Karlqvist, J., and L. Alinde Sundbeck. *Challenges of Using Autonomous Drive Technology for Autonomous Transports in Car Manufacturing*. Master's thesis. Department of Product and Production Development, Division of Production Systems, Chalmers University of Technology, Gothenburg, Sweden, 2016
5. Heutger, M., and M. Kückelhaus. *Self-driving Vehicles in Logistics—A DHL Perspective on Implications and Use-cases for the Logistics Industry*. DHL Trend Research, Troisdorf, Germany, 2014.
6. Ullric, G. *Automated Guided Vehicle Systems: A Primer with Practical Applications*. Springer-Verlag, Berlin Heidelberg, Germany, 2015.
7. Bergerman, M., S. M. Maeta, J. Zhang, G. M. Freitas, B. Hamner, S. Singh, and G. Kantor. Robot Farmers: Autonomous Orchard Vehicles Help Tree Fruit Production. *IEEE Robotics & Automation Magazine*, Vol. 22, No. 1, 2015, pp. 54–63.
8. EETimes. *Nissan Automates Car Towing: Learns about Autonomous Vehicles*, 2016. http://www.automotive-eetimes.com/news/nissan-automates-car-towing-learns-about-autonomous-vehicles. Accessed December 13, 2016.
9. Sein, M. K., O. Henfridsson, S. Purao, M. Rossi, and R. Lindgren. Action Design Research. *MIS Quarterly*, Vol. 35, No. 2, 2011, pp. 37–56.
10. Pan, S. L., and B. Tan. Demystifying Case Research: A Structured-pragmatic-situational (SPS) Approach to Conducting Case Studies. *Information and Organization*, Vol. 21, No. 3, 2011, pp. 161–176.
11. Krueger, R. *Focus Groups: A Practical Guide for Applied Research*. SAGE Publications, Thousand Oaks, CA, 2014.
12. Hjalmarsson-Jordanius, A., N. Sundin, M. Romell, J. Isacson, and C.-J. Aldén. Disrupting Automotive Logistics through a Combined Intelligent and Autonomous Transport Solution. Technical Paper. *Proc., ITS 2017 World Congress*, 29 October–2 November, Montreal, Canada, 2017
13. Simoudis, E. *The Big Data Opportunity in Our Driverless Future*. Corporate Innovators, LLC, Menlo Park, CA, 2017.
14. Hofmann, M. R. Industry 4.0 and the Current Status as well as Future Prospects on Logistics. *Computers in Industry*, Vol. 89, 2017, pp. 23–34.

The Standing Committee on Strategic Management (ABC10) peer-reviewed this paper (18-04607).

Article

TRR
JOURNAL OF THE TRANSPORTATION RESEARCH BOARD

Transportation Research Record
2018, Vol. 2672(7) 34–46
© National Academy of Sciences:
Transportation Research Board 2018
Article reuse guidelines:
sagepub.com/journals-permissions
DOI: 10.1177/0361198118796025
journals.sagepub.com/home/trr
SAGE

Assessing the Impact of App-Based Ride Share Systems in an Urban Context: Findings from Austin

Anurag Komanduri[1], Zeina Wafa[2], Kimon Proussaloglou[3], and Simon Jacobs[4]

Abstract

App-driven ridesharing platforms are gaining popularity and are transforming urban movement patterns in cities throughout the world. Because of privacy and business considerations, their owners have released little information about riders' trip-making characteristics. This lack of data prevents planners and modelers from understanding and quantifying the impact of these new modes on regional travel patterns. In 2016, RideAustin, a not-for-profit company, was established to provide mobility-on-demand services in the Austin region. RideAustin released its dataset of over one million trips to researchers to support transportation planning through a better understanding of urban travel flows. This paper presents findings from an in-depth analysis of this dataset and summarizes key aspects of interest to the transportation research community such as the number of riders, drivers, and trips; total vehicle miles including deadhead miles; and terminal times. The paper also presents findings from two case studies that show the competitiveness of RideAustin versus transit and the utilization of the RideAustin system during the South by Southwest festival. While some of the metrics cannot be readily transferred to other regions, several findings can be used by planners and modelers as they integrate rideshare systems within their planning and modeling frameworks. We also believe that some of the research findings may provide insights into a future system of autonomous and shared vehicles.

Car-sharing in the United States has fluctuated in response to socioeconomic conditions, congestion, and incentive programs (*1*). The most recent rise in car-sharing and ridesharing has been driven in part by congestion in large urban areas and improvements in technology that make ridematching far more efficient and dynamic. This latest round of innovation started with the development of car-sharing systems in Portland in 1997 (*2*), continuing with the development of more subscription-based car-sharing systems such as Zipcar (2000) and Flexcar (2000), the advent of app-based ride-sharing systems such as Uber (2009) and Lyft (2012), and several peer-to-peer car-sharing services.

App-based ridesharing systems have gained significant traction in large urban areas and have generated acceptance among users. In June 2017, Uber announced the completion of five billion rides on its system across six continents and over 450 cities (*3*). The growth rate in the previous 18 months had been impressive. Uber had completed one billion rides by December 2015, two billion rides by June 2016, and then an additional three billion rides from June 2016 to July 2017 (*4, 5*). Collectively, the two largest app-based systems in the U.S. completed about 800 million rides in 2016 in the U.S. alone (*6*) with a significant year-to-year growth rate.

In 2014, the online blog fivethirtyeight.com submitted a successful Freedom of Information Law request for the release of data on for-hire vehicles (FHV) in New York City. The blog subsequently published a series of articles on the impact of FHVs on taxi and transit utilization in the city (*7–10*). This dataset has also been analyzed by researchers in the transportation community (*11–13*). More recently, the San Francisco County Transportation Authority estimated the activity by transportation network companies such as Uber and Lyft in San Francisco (*14, 15*). This report suggests that usage of Uber and Lyft ranges from 129,000 trips on Sundays to 225,000 trips on

[1]Cambridge Systematics, Inc., Los Angeles, CA
[2]Cambridge Systematics, Inc., Austin, TX
[3]Cambridge Systematics, Inc., Chicago, IL
[4]Cambridge Systematics, Inc., Medford, MA

Corresponding Author:
Address correspondence to Anurag Komanduri: akomanduri@camsys.com

Fridays, with travel on these modes representing about 15% of all vehicle trips within San Francisco.

These findings indicate that the ridesharing ecosystem is here to stay and is likely to accelerate as more systems come online, as rideshare offerings become more consumer-driven (pooled trips, commuter lines, multilingual drivers, and services compliant with the Americans with Disabilities Act), and as rideshare systems become increasingly used for first-and-last mile services. Therefore, the subject is of interest and there is a great need to study data from ridesharing systems to understand better how the systems operate in urban areas, how they compete with or complement existing transit systems, what types of trips are served, and rider and driver profiles. However, planners have been hampered by the lack of readily available data that can be studied in detail since ridesharing companies consider trip data as proprietary information.

This paper addresses this gap by analyzing and documenting findings from a dataset released by a ridesharing company in Austin, Texas, which provides detailed travel information. We believe that this is the first open dataset of its kind released in the U.S. offering a blueprint for the variables that are available for detailed analysis. Based on these insights, agencies may request synthesized data from rideshare companies to support city and regional planning efforts.

The paper first describes the dataset that was released for research and outlines the steps we took prior to analyzing the data. A brief discussion follows about how the data may be used to build datasets that traditionally support models, understand system performance, and perform case studies. Finally, we document our conclusions and identify key metrics that cities and urban planners can request as they develop data-sharing agreements with ridesharing companies that operate in their regions.

Background and Data Description

In May 2016, established ridesharing companies such as Uber and Lyft ceased to operate in the City of Austin in response to Prop 1, a measure which required mandatory fingerprinting for drivers operating in the city (*16*). To fill the void left by the larger ridesharing companies, several smaller ridesharing companies and start-ups established new compliant services including Fare, Fasten, Get Me, InstaRyde, Wingz, zTrip, and RideAustin (*17*).

Since RideAustin was founded as a not-for-profit business with the explicit motto of serving the greater Austin region, it is expected that an overwhelming majority of its riders are local residents. One of the founding principles of the company is to provide open data to promote transparency and support research (*18*). RideAustin started operations in mid June 2016 and has served nearly two million trips to date in the greater Austin region. In May 2017, RideAustin released to the research community a comprehensive dataset of 1.5 million trips from June 2016 to mid April 2017 (*19*). To the best of our knowledge, all trips undertaken on the RideAustin system are included in this database, which presents a complete picture of system utilization without the presence of sampling biases. However, since this database represents data from one ridesharing company, there may be some self-selection bias which is hard to determine unless data from all ridesharing companies were also made available.

Figure 1 describes the trip-making patterns by month and day of the week. The RideAustin system experienced continuous growth from June 2016 through March 2017 except in the months of November and December, when total trips dipped compared with October. The majority of trips took place on weekends (Friday to Sunday) (60%) as opposed to other days of the week.

The dataset includes standard information about the start and end times and locations of each trip. (Latitudes and longitudes are truncated to the third decimal point, resulting in location accuracy of about 110 meters. This limits the sharing or revealing of any personal identifying information.) Also recorded are distance and fare, type of vehicle used, and time when the app was activated for the ride. Similar information has been available for ridesharing and taxi data released in other cities such as New York and Chicago. Where this dataset is different, and where it provides more value, is that it also provides a consistent driver and rider identifier (ID). This allows research into rider and driver behavior and building key performance metrics such as deadhead miles and terminal times, aggregating rider trips to tours, and segmenting the RideAustin customers and drivers.

The RideAustin database covers a period of about 10 months and translates to about 5,000 trips per day (Figure 1). The highest system utilization on any given day was about 20,600 trips during the South by Southwest (SXSW) festival in March 2017. On that day, the system served the highest number of riders (~14,500) and registered the highest number of active drivers who served at least one trip (~2,200).

The dataset was organized and summarized by customer, by driver, by day, by month, and by time of day to generate key summary statistics described in the paper. Where needed, records were reorganized by customer and by driver to generate chronologically linked trips to develop measures such as time spent between trips and miles traveled between trips. Overall, the research team employed very few data modification strategies, making a conscious effort to use the data in its native format to focus on the underlying value of the data instead of advanced analytical techniques. About 17,000 rides were

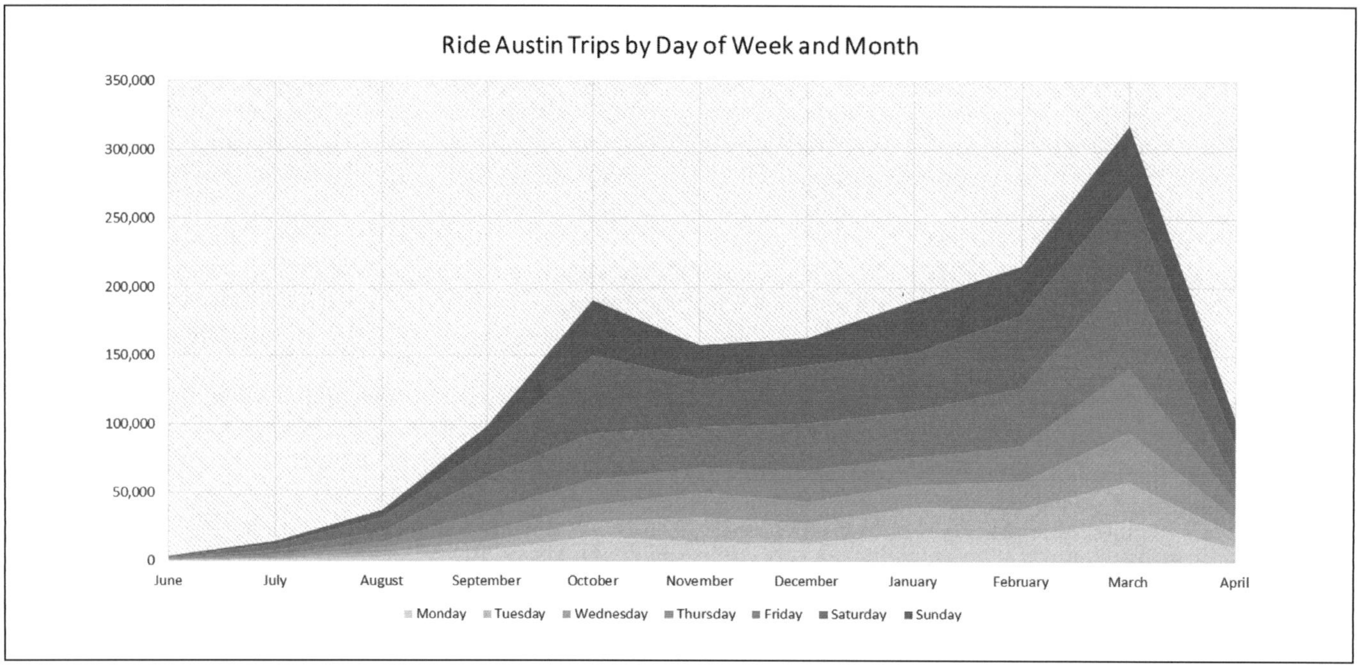

Figure 1. Summary of RideAustin trips over time by day of the week.

dropped from the dataset, however, either because the trip distance was reported to be zero miles or because the average speed for the trip exceeded 60 miles per hour.

Research Findings

The team's research focused on five key areas: (a) developing system-level metrics, (b) outlining users' travel patterns, (c) building trip profiles to support travel demand models, (d) comparing RideAustin service attributes with transit service, and (e) studying a special case of trip-making behavior during the SXSW festival.

System Metrics and Key Performance Indicators

Traditional modeling metrics on total trips, vehicle-miles traveled, total travel times, and fares paid can be readily developed using the RideAustin database. In addition, the structure of the database allowed us to calculate key variables such as deadheading, terminal times, and the number of active riders and drivers.

Deadheading Time and Distance. Since the database tracks the ID of both the driver and rider, it was possible to build a database of linked trips for both drivers and passengers. For the purposes of this paper, we focused primarily on linked driver trips since they allow us to understand time and distance spent deadheading, that is, non-revenue miles. In calculating this metric, we assumed that at the start and end of every shift, drivers are expected to travel about two miles and for five minutes to reach their first trip origin and from their last destination. This metric reflects the average distance of 2.15 miles traveled by drivers on non-revenue legs in the middle of a shift. We then assumed an average speed of about 25 miles per hour to develop the metric for travel time.

In cases where drivers spent more than 30 minutes between a drop-off and a subsequent pick-up, we assumed that the "start/end of shift" rules apply. (This may be a conservative estimate and could result in longer deadheading times. It can always be refined if industry standards reveal a more accurate upper limit for wait times observed for drivers between trips.) In calculating the deadhead distance, we applied a straight-line distance between the end of the previous trip and the start of the next trip. Straight-line distance is the lowest measure of deadheading distance since many drivers very likely drive idly on streets before picking up their next fare or follow network routes that are longer than the straight-line distance. The objective of this metric is merely to measure the "floor" on deadheading distance. Therefore, the estimates presented here most likely represent the lower end of the actual deadhead distances.

Terminal Time. The database presents a series of time stamps including the initialization of the app to call for a ride, the driver's acceptance, the driver's arrival at the trip start location, the actual trip start time, and the time when the trip ended. Terminal time was measured as the time spent between the initialization of the app and the trip start time. It can be viewed as the "wait time" for the door-to-door RideAustin service.

Results. Table 1 summarizes key statistics from the RideAustin database including the following:

- All system-level activity metrics and key performance indicators at the individual level grew over time, suggesting the greater appeal and increasing use of the system. The exceptions were the deadheading distances and travel times, which gradually reduced, suggesting that the system operated more efficiently over time.
- The first four months of operation attracted relatively few drivers to the system. Starting in October, the number of unique drivers who offered at least one ride per month was over 2,000, growing to above 2,600 in 2017. The average monthly trips offered by a driver remained over 60 (two or more trips per day) since September 2016.
- The average trip distance remained relatively stable at between 2.5 and 3.5 miles suggesting that the system is used for relatively short trips. Furthermore, the average trip fare varied from \$11.80 to \$15.50 per month which means riders paid between \$3.5 and \$4.6 per mile (or more than a dollar a minute of travel time). This indicates a rather high willingness-to-pay among riders, especially when compared with the \$14.7 per hour value of time in the CAMPO travel demand model. (CAMPO is the Capital Area Metropolitan Planning Organization in Texas, which includes Bastrop, Burnet, Caldwell, Hays, Travis, and Williamson Counties.)
- The total number of unique riders also increased from October 2016 onwards with over 60,000 unique riders in October 2016 and February 2017, reaching a high of 85,000 unique riders in March 2017. March coincides with the SXSW festival, while October includes football games at the University of Texas as well as the Austin City Limits music festival. It is possible that the closure of the university for parts of November and December for holidays has an effect on utilization.
- The average terminal times at the inception phase of the system were longer than nine minutes. However, as the number of drivers grew and the system operations stabilized, the average terminal times trended downwards to about six minutes. Terminal time as a percentage of total terminal and trip time was between 31% and 33% between November and April. This important finding suggests riders' "willingness to wait" in exchange for a ride with customized pick-up and drop-off locations.
- In total, RideAustin drivers drove at least 8.7 million vehicle miles over 10 months. About 37% of these total miles were deadheading miles. This important finding provides a measure of how much extra travel is contributed by ridesharing systems. When we examined deadhead travel times, we calculated that drivers spend just over 50% of their total travel time deadheading.

Customer Usage Profiles

In total, nearly 260,000 unique rider IDs were found in the RideAustin database. Research conducted by Deloitte (*20*) suggests that the smartphone replacement cycle tends to start after one year and peaks at about 18 months. A survey conducted by Statista (*21*) found that more than half the respondents replaced their phones after two years, while the World Bank (*22*) found that there were 127 mobile cellular subscriptions per 100 people in the U.S. Therefore, it is likely that there may be some duplication in identifying customers since some customers have multiple phones, have replaced their phone's operating system (OS to Android), or have obtained a replacement phone or telephone number. However, switching phones or numbers may not always result in a new customer ID if more stable emails or app-download accounts are used to link to IDs. For the purposes of this paper, we assumed that each unique ID represents a unique customer.

The total population of Austin Metro is about two million while the City of Austin has about 910,000 residents. RideAustin is focused on mostly serving local residents. (Preliminary analysis of RideAustin data from during the SXSW festival suggests that the system was used by some out-of-town visitors but it is unclear whether there is a large share of out-of-town visitors during other seasons.) This suggests that between 10% and 25% of the regional population has made at least one trip on the system, a remarkably high adoption rate in a very limited period of time.

The data also suggest a very low retention rate among riders who use the system, however (Table 2). Nearly 110,000 riders used the system for one day only while about 175,000 riders used the system over a duration of five days or less. Only about 50,000 riders hailed rides more than 60 days apart. In contrast, over 60% of users of transit in Austin reported using the system for six months or more (*23*), suggesting that there is considerably lower retention among RideAustin riders than among transit users in the region. Repeat riders who used RideAustin at least 90 days apart (between their first and last trips) made the most trips. Although this

Table 1. Key Monthly System Statistics and Key Performance Indicators

	June	July	August	September	October	November	December	January	February	March	April
System-level travel statistics											
Trips	3,505	13,376	36,174	95,590	186,251	156,476	161,759	189,045	214,703	316,640	104,131
Unique riders	1,879	6,247	16,114	37,256	61,631	48,689	51,057	57,573	62,333	85,585	40,089
Unique drivers	229	463	793	1,528	2,147	2,089	2,340	2,684	2,799	3,558	2,849
Trip distance (miles)	9,039	43,967	125,575	332,370	675,964	569,355	600,725	709,920	816,012	1,166,124	389,435
Trip time (min)	41,101	164,859	460,728	1,267,390	2,531,223	2,052,145	2,066,641	2,365,083	2,749,204	4,340,573	1,387,048
Trip cost ($)	$42,289	$187,815	$528,133	$1,289,227	$2,747,964	$2,036,736	$2,092,686	$2,497,063	$2,843,031	$4,922,204	$1,484,371
Terminal time (min)	31,328	122,676	295,419	760,064	1,325,353	1,030,721	1,010,828	1,075,864	1,271,157	1,958,631	637,858
Deadhead distance (miles)	6,759	30,517	83,429	208,717	393,802	329,859	347,393	405,030	456,739	634,643	216,874
Deadhead time (min)	46,889	195,116	518,572	1,362,094	2,588,350	2,152,088	2,281,789	2,709,552	2,959,419	4,228,792	1,411,321
Individual-level key performance indicators											
Trips per unique rider	1.87	2.14	2.24	2.57	3.02	3.21	3.17	3.28	3.44	3.70	2.60
Trips per unique driver	15.31	28.89	45.62	62.56	86.75	74.90	69.13	70.43	76.71	88.99	36.55
Average trip distance	2.58	3.29	3.47	3.48	3.63	3.64	3.71	3.76	3.80	3.68	3.74
Average trip duration	11.73	12.32	12.74	13.26	13.59	13.11	12.78	12.51	12.80	13.71	13.32
Average trip fare	12.07	14.04	14.60	13.49	14.75	13.02	12.94	13.21	13.24	15.55	14.25
Average terminal time	8.94	9.17	8.17	7.95	7.12	6.59	6.25	5.69	5.92	6.19	6.13
Terminal time %	43.3%	42.7%	39.1%	37.5%	34.4%	33.4%	32.8%	31.3%	31.6%	31.1%	31.5%
Deadhead time %	53.3%	54.2%	53.0%	51.8%	50.6%	51.2%	52.5%	53.4%	51.8%	49.3%	50.4%
Deadhead distance %	42.8%	41.0%	39.9%	38.6%	36.8%	36.7%	36.6%	36.3%	35.9%	35.2%	35.8%

Table 2. Customer Segments Based on First Ride and Last Ride Metrics

Days between first and last trip	Riders	Trips	Trips/riders
1 day (same day)	109,242	134,879	1.23
2 days	29,303	84,413	2.88
3 days	19,204	79,735	4.15
4 days	11,947	61,922	5.18
5 days	5,718	33,629	5.88
6–10 days	8,094	48,703	6.02
11–15 days	2,885	13,415	4.65
15–30 days	7,531	38,822	5.15
31–60 days	11,546	75,435	6.53
61–90 days	10,323	92,021	8.91
Over 90 days	43,778	814,676	18.61
Total	259,571	1,477,650	5.69

segment comprises only 17% of total riders, they made over 55% of all trips, or almost 19 trips per rider.

Additional analyses revealed that over 70% of riders made one to five trips during the analysis period while another 16% made five to 10 trips. Only about 600 riders made more than 100 trips, and of these only 84 made at least 200 trips, suggesting a relatively modest number of high-frequency riders. These findings suggest that RideAustin tends to serve a large number of individuals, but the use is focused on specific times of need with low customer retention.

Table 3 shows the customer retention statistics (called "churn" in market terms) by month. It suggests that 28.6% of the June 2016 "early adopters" continued to use the system through April 2017. When examining the metrics from a market size perspective, over 4,000 riders who used RideAustin for the first time in September and October 2016 continued to use the system in April 2017. Similarly, over 65% of the riders remained active for one month or less.

Riders who made at least 20 trips (14,278 unique riders) were analyzed further to examine their trip ends. The most common trip end location, with at least three trip ends, was defined as a primary location. Similarly, the next most common trip end location was assigned as a secondary location. Trip ends were then marked as primary, secondary, or "other":

- There is an interesting and expected trend in travel patterns by time of day. The primary location tends to be a starting point for the majority of trips starting between 6:00 a.m. and noon. Almost 50% of all travel in the 11:00 p.m. to 6:00 a.m. period ends at a primary location. Such trip start patterns are also observed commonly for home locations in a traditional trip analysis framework.
- The origin share for secondary locations holds steady throughout the day, indicating that the place type associated with a secondary location most likely varies from rider to rider.
- Nearly 45% of trip starts and 51% of trip ends made by riders arrived at less common destinations labeled as "other."
- Overall, the flows between these trip ends is imbalanced. While 38% of trip origins are at the primary location, only 30% of trip destinations are at the primary location, suggesting that people tend to use RideAustin more to depart from rather than to arrive at their primary location.

Metrics to Support Travel Demand Models

Since the RideAustin dataset does not explicitly provide rider demographics or purpose of travel, we focused on developing two metrics that traditionally support the validation of travel demand models and are presented below.

Trip Length Distribution. This metric is commonly used to validate trip distribution or destination choice models and was developed by using the trip start and trip end locations (Figure 2). On any given day, over 80% of all RideAustin trips were shorter than six miles, suggesting that the service serves primarily local movements. There is a significant difference in average trip lengths by day of the week. Weekend trips are shorter than weekday trips indicating that weekend travel is used for hyperlocal travel.

Travel by Time of Day. The time-of-day travel patterns are skewed towards the evening hours (6:00–11:00 p.m.) and night time (11:00 p.m.–2:00 a.m.) in Figure 3. In fact, nearly 60% of all RideAustin trips are made between 6:00 p.m. and 6:00 a.m. This suggests that the system is primarily used for recreational and leisure activities that tend to spike during this time period. On weekends, the share of travel in the 2:00–6:00 a.m. time period is double the share of the weekday travel, suggesting the greater use of car-sharing for late-night recreational weekend travel.

When examining the 10-month trend in utilization in Figure 3, we observe that the market share for travel is growing during the commute time periods (6:00–9:00 a.m. and 3:00–6:00 p.m.) and during the middle of the day. This indicates a gradually greater acceptance of RideAustin as a reliable mode of transportation to support repetitive and time constrained travel.

Table 3. Customer Retention using First-and-Last Ride Metrics

		2016							2017				
	Start month/ end month	June	July	August	September	October	November	December	January	February	March	April (half month)	Total
2016	June	502	65	46	42	76	59	60	77	101	315	536	1,879
	July		1,904	212	171	244	128	189	222	307	944	1,161	5,482
	August			5,495	541	654	383	534	530	685	1,983	2,509	13,314
	September				15,297	3,153	888	949	948	1,268	3,376	3,790	29,669
	October					29,416	1,548	1,221	1,285	1,721	4,220	4,261	43,672
	November						18,144	1,476	1,132	1,237	2,873	3,080	27,942
	December							15,079	2,494	1,384	3,201	2,734	24,892
2017	January								19,338	1,820	3,013	2,877	27,048
	February									23,333	3,229	2,388	28,950
	March										39,970	4,384	44,354
	April (half month)											12,369	12,369
	Total	502	1,969	5,753	16,051	33,543	21,150	19,508	26,026	31,856	63,124	40,089	259,571

Duration between riders' first and last trips	Number of riders	Share of total riders
<1 month	168,478	65%
1–2 months	14,538	6%
2–3 months	8,509	3%
3–4 months	7,341	3%
4–5 months	6,280	2%
5–6 months	6,266	2%
6–7 months	4,343	1.5%
7–8 months	2,367	1%
8–9 months	1,045	<0.5%
9–10 months	315	<0.5%
Current active riders	40,089	15%
Total riders	259,571	

Table 4. Start and End Time Distribution by Trip End Type and by Time of Day

Time of day	Primary origin share	Primary destination share	Secondary origin share	Secondary destination share	Other origin share	Other destination share
Morning peak (6–9 a.m.)	62%	20%	19%	38%	19%	42%
Late morning (9–12 a.m.)	56%	21%	18%	29%	27%	50%
Mid-day (12 a.m.–3 p.m.)	46%	23%	17%	20%	37%	57%
Afternoon peak (3–6 p.m.)	42%	27%	20%	18%	38%	55%
Evening (6–11 p.m.)	40%	26%	16%	14%	44%	60%
Night (11 p.m.–2 a.m.)	16%	45%	15%	17%	70%	38%
Overnight (2–6 a.m.)	18%	48%	17%	19%	65%	33%
Daily	38%	30%	17%	19%	45%	51%

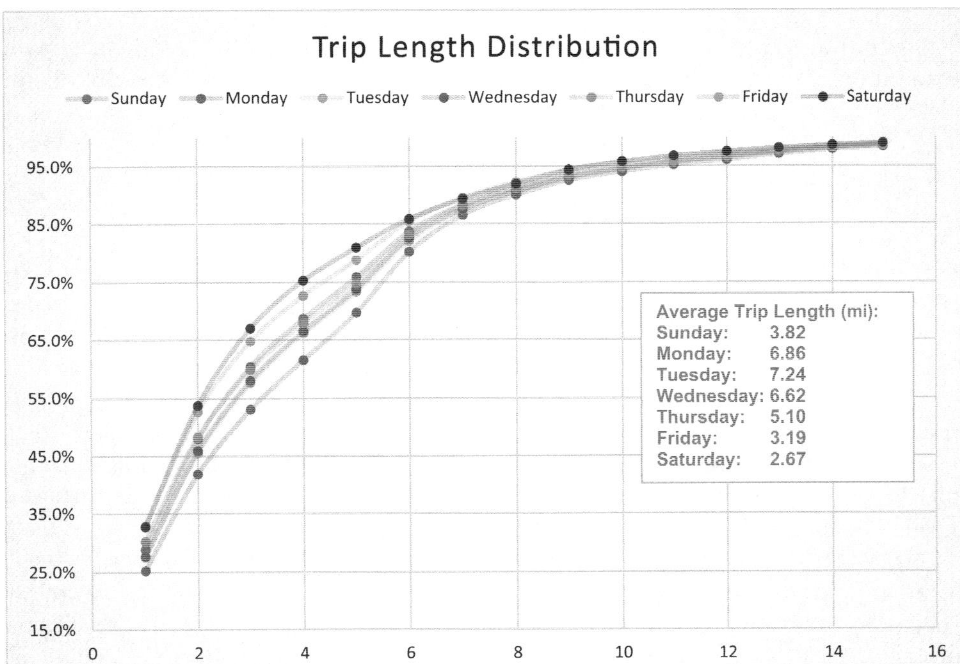

Figure 2. Trip length distribution by day of the week.

Transit Competitiveness Case Study

There is anecdotal evidence which suggests that ridesharing systems may have adversely affected transit ridership in urban areas. While the convenience of door-to-door service provided by ridesharing systems cannot be matched at a systemwide level by transit, we summarized and contrasted quantitative metrics such as in-vehicle travel times and out-of-vehicle travel times by generating transit levels of service for trips made on the RideAustin platform. (A quick review indicated that transit in Austin is almost always cheaper than making a trip using rideshare services, due to high flagging costs (fewer than 150 RideAustin trips had fares lower than $3.75). Clearly fare competitiveness is not what is driving the use of

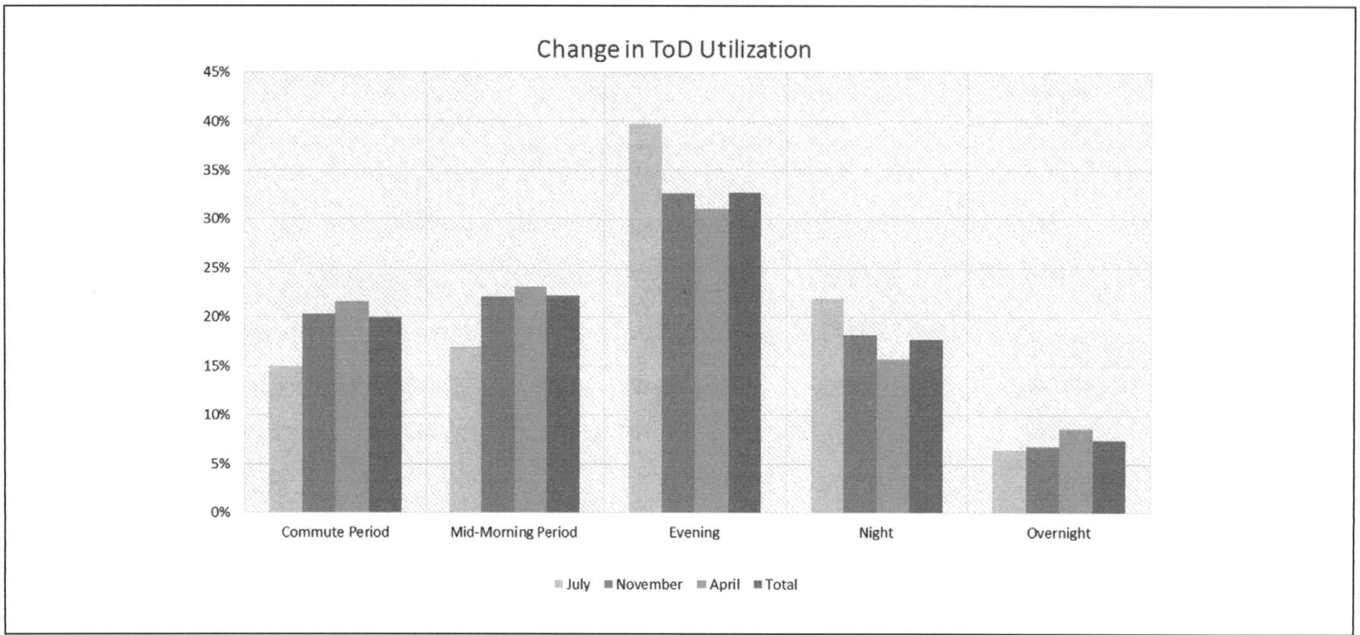

Figure 3. Temporal distribution of RideAustin trips by time of day and month.

rideshare systems.) Other transit competitiveness measures found in the literature include service frequency (*24*, *25*), service reliability (*26*), transit accessibility (*24*), parking availability and cost (*24*, *26*), number of transfers needed (*25*), toll pricing (*26*), land use density (*26*, *27*), transit priority treatment (*27*), pedestrian environment (*27*), and ride smoothness (*27*). We limited our analysis to frequency, number of transfers, and travel times, which we could readily utilize using the RideAustin dataset and transit path-building algorithms.

Capital Metropolitan Transportation Authority's (CapMetro) General Transit Feed Specification (GTFS) system was mapped onto OpenTripPlanner, a Java-based multimodal trip planning engine. It builds a mathematical graph representation of a metropolitan area from a transit schedule in GTFS format and street network data from OpenStreetMap. Given a rider's origin, destination, timing, and desired mode request, OpenTripPlanner performs a graph search to create itineraries that fulfill the request, or an error if no itinerary can be found. We wrote a Python script to convert origin–destination data into OpenTripPlanner requests, sent the requests to OpenTripPlanner, and processed the returned itinerary to determine whether a transit trip was found and what the characteristics of that trip were. The trip planner was subsequently queried for all trips made on the RideAustin system in October 2016 to generate corresponding transit trips. Table 5 shows the total number of completed RideAustin trips in October 2016 by time of day, transit availability, and the number of transfers required to ride transit for each of these trips:

- More than 60% of these trips could have been provided by a single-seat transit ride (118,557 trips) suggesting that many of RideAustin trips occur in CapMetro's core service area.
- The absence of transit paths suggests trips outside CapMetro's service area. Only 11% of all trips made on RideAustin had no possible transit path (21,407 trips). An additional 3% of rides required two or more transfers – a market where transit typically cannot attract non-captive riders (4,833 trips).
- Among the rides that have a transit path, nearly 39,000 trips started at least 15 minutes after the start time for the RideAustin trip while nearly 6,000 trips started at least an hour later. (We felt potential riders would not want to wait longer for a ride than their in-vehicle time, hence we chose 15 minutes because the average RideAustin trip time is about 13 minutes.) It is possible that quite a few RideAustin riders chose not to use transit due to its inconvenient schedule.
- However, there were still about 125,500 trips (67% of total) where transit provides a service when and where people need the ride.

We then focused on those 125,500 RideAustin trips that had a transit path and whose transit trip start time was within 15 minutes of the start time of their RideAustin trip (Table 5). We calculated the total travel time by transit by adding in-vehicle and out-of-vehicle travel time, which includes walk time and wait time. We should note that in the case of the RideAustin trips, the

Table 5. Transit Competitiveness Measures

Time of day	No transit path	Zero transfer transit path	1 transfer transit path	2+ transfers transit path	Total trips
Morning peak	1,185	4,414	2,516	369	8,484
Late morning	2,495	10,400	5,042	688	18,625
Mid-day	2,919	15,532	7,050	798	26,299
Afternoon peak	3,466	16,387	6,823	783	27,459
Evening	5,578	38,471	11,956	1,305	57,310
Night	4,114	24,890	5,482	556	35,042
Overnight	1,650	8,463	2,585	334	13,032
Total trips	21,407	118,557	41,454	4,833	186,251
Percentage	11%	64%	22%	3%	
Transit start time is more than 15 mins after RideAustin start time	N/A	25,681	11,532	1,986	39,199

Metrics	Zero transfer transit path	1 transfer transit path	2+ transfers transit path	Total transit available trips
Transit total time less than RideAustin	1,112	94	3	1,209
Transit total time 100–120% of RideAustin	1,226	152	5	1,383
Transit total time 120–140% of RideAustin	2,091	307	12	2,410
Transit total time >140% of RideAustin	88,447	29,369	2,827	120.643
Total	92,876	29,922	2,847	125,645
Transit IVTT less than RideAustin IVTT	1,819	154	2	1,975
Transit IVTT 100–120% of RideAustin IVTT	1,679	162	12	1,853
Transit IVTT 120–140% of RideAustin IVTT	2,699	352	7	3,058
Transit IVTT >140% of RideAustin IVTT	88,447	29,254	2,826	120,527
Total	92,876	29,922	2,847	125,645
Transit OVTT less than RideAustin OVTT	2,942	673	39	3,654
Transit OVTT 100–120% of RideAustin OVTT	1,899	532	30	2,461
Transit OVTT 120–140% of RideAustin OVTT	2,503	684	54	3,241
Transit OVTT >140% of RideAustin OVTT	85,532	28,033	2,724	116,289
Total	92,876	29,922	2,847	125,645

Note: IVTT = in-vehicle travel time; OVTT = out-of-vehicle travel time.

out-of-vehicle travel time is limited to the terminal time between the time the request was made and the time a passenger was picked up:

- Most trips in the database confirm the advantage of the on-demand door-to-door nature of a ride-share service. There were only about 1,200 trips where transit has a better overall time than the RideAustin service. In fact, a majority of transit trips were slower by 40% or more when compared with RideAustin trips.
- This finding along with the summary statistics presented in Table 1 suggests that RideAustin riders have a high value of time and are willing to pay a premium to minimize their travel time.

South by Southwest (SXSW) Case Study

SXSW is an annual festival of film, music, and technology that takes place in Austin, Texas, in March. The event typically spans 10 days, generates a lot of recreational activity in the downtown area, and strains the limited parking availability since its attendance in 2017 was over 70,000 attendees (*28*). We analyzed the database to examine whether visitors to the festival used RideAustin. By contrasting the patterns of use during the festival with a "typical" period we assessed to what extent RideAustin was being used by residents alone or by visitors to the region as well.

To quantify the impact of the festival on RideAustin metrics, we analyzed two distinct periods including the dates of the SXSW festival (March 10 to March 19) and a period of similar length after the festival ended (March 31 to April 9). To study out-of-region visitor travel, all hotel locations were geocoded (accessed using a web-based application program interface). In addition, Airbnb locations that were booked during the two study periods were also geocoded (accessed through an open data system forum on a website called insideairbnb.com). Some key findings are identified below:

Table 6. RideAustin Trips during SXSW and after SXSW

Study period	Ride Austin trips	Trips near airport		Trips near hotels		Trips near Airbnb locations		Trips near convention center	
		Start	End	Start	End	Start	End	Start	End
Mar 10–Mar 19	166,055	5,690	9,325	55,800	62,618	21,858	20,408	6,520	10,547
Mar 31–Apr 9	95,834	4,023	5,861	26,811	27,985	13,418	13,418	2,432	2,943
Percent increase	73%	41%	59%	108%	124%	63%	52%	168%	258%

- The total number of RideAustin trips during the festival period was about 166,000 compared with only 95,800 trips during the typical March 31 to April 9 period (57% of SXSW trips) suggesting that the festival attracted a significantly large number of RideAustin trips.
- It is important to note that that the number of RideAustin trips has continued to increase over time as the system keeps growing and riders familiarize themselves with it. Therefore, the observed trips during the typical March 31 to April 9 period should not reflect any slowdown in RideAustin utilization.
- Festival attendees booked about 8,000 Airbnb locations for at least one day during the festival and could also stay at the 190 hotels in the Austin area, hosting nearly 26,000 rooms. In April, the number of booked Airbnb locations dropped to about 6,500 (18% fewer than during SXSW).
- The RideAustin trips during both time periods were mapped using a 200-meter buffer around four types of key activity centers including hotels, booked Airbnb locations, the airport, and the convention center. These four types of locations attract a large portion of SXSW attendees and could serve as a surrogate measure for RideAustin trips headed to/from the festival.
- As shown in Table 6, there was a significant increase in activity within 200 meters of the airport, hotels, booked Airbnb locations, and the convention center during SXSW:
  - RideAustin trips with a destination near the convention center (a major venue for SXSW) were triple the number of trips during a typical day, while trips originating at the convention center were more than double the typical observed activity.
  - Trips that start or end near the hotels were double the typical RideAustin activity near the same hotels. Similarly, trips to/from Airbnb locations were 50% higher than a typical day.
  - Finally, RideAustin trips to and from the airport were 60% and 40% more frequent, respectively, compared with a typical time period in Austin.
  - Across all four location types, the growth in RideAustin use was significant and greater than the typical monthly growth. These patterns clearly suggest that non-residents attending SXSW were heavy users of the RideAustin app.

Conclusions and Future Research

This paper takes a first step towards quantifying the travel patterns of both riders and drivers in a ridesharing context using a unique dataset that has yielded the following important findings:

- The RideAustin data allow the *calculation of deadheading times and distances* which indicate a significant amount of deadheading travel as a percentage of total travel. For city officials who want to evaluate the effect of ridesharing systems on regional congestion, this analysis provides data and an approach to assess the degree of additional deadheading travel due to such systems.
- Riders *are willing to wait an average of six minutes* for RideAustin services. These terminal times can be incorporated in any rideshare mode in a modeling context. They also provide useful metrics for transit agencies interested in designing high frequency routes.
- Understanding the travel patterns of ridesharing riders could help a transit agency *design routes that serve key markets directly*. Transit agencies could also decide to truncate or eliminate poorly performing routes that may be better served by more flexible ridesharing.
- RideAustin riders *have a high willingness-to-pay for a premium service*. However, a relatively low rider retention rate suggests that most riders choose this option only when no other good options are available, a challenge for smaller ridesharing systems.
- Finally, this research *proves the value of using ridesharing trip data to support transportation planning*.

While data on driver ID may be masked to prevent identification, retaining an ID can help quantify deadheading. Masking of trip ends with more aggregate geographic measures maintains privacy while not adversely affecting the dataset quality.

- Cities *looking to understand the impact of ridesharing* on their transportation system should seek information that, at the very minimum, provides the following details:
  - Number of unique customers served, along with a frequency of customers who make different numbers of trips to understand system utilization.
  - Trip tables broken down by time period, day of the week, and month of the year to align transit routes to match demand better. True trip origins and destinations may be masked using aggregation methods.
  - Total mileage and travel times reported by drivers when using the app split by revenue and non-revenue trips. If possible, such mileage should be reported by time of day, day of the week, and by geography to measure vehicle-miles (times) traveled and deadheading miles (times).
  - Data sharing should be discussed in parallel with data confidentiality arguments so that businesses feel comfortable sharing the data for greater social good without adversely affecting their revenue.

There are many areas of future research that can be pursued using this rich travel dataset. This list is not intended to be comprehensive but suggests research opportunities using similar datasets:

- In-depth assessments of regular customers to understand their usage patterns
- Segmentation to understand early adopters of technologies such as autonomous vehicles
- Economic analysis to understand driver earnings after accounting for vehicle depreciation, fuel costs, and possibly tolls
- Research to right-size fleets to meet customer demand and quantify the number of cars, and deadheading that is likely to occur under a fully autonomous fully ridesharing future
- Examination of primary and secondary locations for riders by employing a time-of-day dimension to identify home locations
- Assignment of broad demographic profiles to better understand system users
- Development of tour profiles for riders and drivers to better understand system utilization

Acknowledgments

The authors acknowledge the generosity of RideAustin in releasing their rich trip dataset to a curated environment hosted by Datadotworld. We also acknowledge the contributions of Samuel Van Hecke and Cemal Ayvalik who helped shape the analysis.

Author Contributions

The authors confirm contribution to the paper as follows: study conception and design: Anurag Komanduri: data collection, Anurag Komanduri; analysis and interpretation of results: Anurag Komanduri, Kimon Proussaloglou, Zeina Wafa; draft manuscript preparation: Anurag Komanduri, Kimon Proussaloglou. All authors reviewed the results and approved the final version of the manuscript.

References

1. Chan, N. D., and S. A. Shaheen. Ridesharing in North America: Past, Present, and Future. *Transport Reviews*, Vol. 32, No. 1, 2012, pp. 93–112.
2. *A Brief History of Car Sharing*. Portland Bureau of Transportation, City of Portland, 2011. https://www.portlandoregon.gov/transportation/article/370287. Accessed July 2, 2017.
3. Holt, R., A. Macdonald, and P. D. Gore-Coty. 5 Billion Trips, *Uber Newsroom Press Release*, June 29, 2017. https://newsroom.uber.com/5billion/. Accessed July 2, 2017.
4. Kokalitcheva, K. Uber Completes 1 Billion Rides, *FORTUNE*, December 31, 2015. http://fortune.com/2015/12/30/uber-completes-1-billion-rides/. Accessed July 2, 2017.
5. Yan, S. *Uber Hits 2 Billion Trips*, CNN Tech, July 18, 2016. http://money.cnn.com/2016/07/18/technology/uber-two-billion-rides/index.html. Accessed July 2, 2017.
6. Solomon, B. Lyft Rides Tripled Last Year, but Remains Far Behind Uber, *Forbes Magazine*, January 5, 2017. https://www.forbes.com/sites/briansolomon/2017/01/05/lyft-rides-tripled-last-year-but-remains-far-behind-uber/#261cc3fd199e. Accessed July 2, 2017.
7. *FiveThirtyEight GitHub link for-hire vehicle-data*. https://github.com/fivethirtyeight/uber-tlc-foil-response. Accessed July 2, 2017.
8. Silver, N, and R. Fischer-Baum. *Public Transit Should Be Uber's Best Friend*. FiveThirtyEight, August 28, 2015. https://fivethirtyeight.com/features/public-transit-should-be-ubers-new-best-friend/. Accessed July 2, 2017.
9. Carl Bialik, C., A. Flowers, R. Fischer-Baum, and D. Mehta. *Uber is Serving New York's Outer Boroughs More than Taxis Are*, FiveThirtyEight, August 10, 2015. https://fivethirtyeight.com/features/uber-is-serving-new-yorks-outer-boroughs-more-than-taxis-are/. Accessed July 2, 2017.
10. Fischer-Bau, R., and C. Bialik. *Uber is Taking Millions of Rides Away from Taxis*, FiveThirtyEight, October 13, 2015. https://fivethirtyeight.com/features/uber-is-taking-millions-of-manhattan-rides-away-from-taxis/. Accessed July 2, 2017.

11. Correa, D., K. Xie, and K. Ozbay. Exploring the Taxi and Uber Demands in New York City: An Empirical Analysis and Spatial Modeling. Presented at 96th Annual Meeting of the Transportation Research Board, Washington, D.C., 2017.
12. Davidson, A., J. Peters, and C. Brakewood. Interactive Travel Modes: Uber, Transit, and Mobility in New York City. Presented at 96th Annual Meeting of the Transportation Research Board, Washington, D.C., 2017.
13. Viswanathan, K., and A. Komanduri. Evaluating Taxi Data from 2010 to 2015. *Proc., 6th Transportation Research Board's Innovations in Travel Modeling Conference*, Denver, CO, 2015.
14. San Francisco County Transportation Authority. *A Profile of San Francisco Transportation Network Company Activity*, 2017. http://www.sfcta.org/sites/default/files/content/Planning/TNCs/TNCs_Today_112917.pdf. Accessed July 2, 2017.
15. Charlton, B. Visualizing Uber and Lyft trips in San Francisco: More than 200,000 Trips a Day, *HACKERNOON*, June 13, 2017. https://hackernoon.com/visualizing-uber-and-lyft-usage-in-san-francisco-928208b1978a. Accessed July 2, 2017.
16. Prop 1 Fails, Marking Defeat for Uber and Lyft in Austin, *KXAN News*, Austin, TX. http://kxan.com/2016/05/07/prop-1-fails-marking-defeat-for-uber-and-lyft-in-austin/. Accessed July 3, 2017.
17. The Complete List of Ride-Sharing Companies Available in Austin, *CultureMap*, Austin, TX. http://austin.culturemap.com/news/innovation/06-03-16-ride-hailing-ridesharing-options-in-austin-transportation-network-companies-apps/. Accessed July 3, 2017.
18. RideAustin. *A Nonprofit Rideshare Built for Austin*. http://www.rideaustin.com/#a-culture-of-transparency. Accessed July 3, 2017.
19. Comprehensive Rides Data from 06-02-2016 through 04-13-2017, *RideAustin*, Austin, TX. https://data.world/ride-austin/ride-austin-june-6-april-13. Accessed July 3, 2017.
20. One Billion Smart Phone Upgrades. Deloitte Technology, Media & Telecommunications Practices. https://www2.deloitte.com/content/dam/Deloitte/global/Documents/Technology-Media-Telecommunications/gx-tmt-pred15-one-billion-smartphone.pdf. Accessed October 17, 2017.
21. Statista, The Statistics Portal. *How Long Do You Think You Will Own Your Current Mobile Phone from the Date of Purchase Until You Replace/Discard It?* https://www.statista.com/statistics/241036/ownership-time-length-of-current-mobile-phone-until-replacement/. Accessed November 9, 2017.
22. World Bank. *Mobile Cellular Subscriptions*, https://data.worldbank.org/indicator/IT.CEL.SETS.P2. Accessed November 9, 2017.
23. Capital Metro. *Capital Metro Origin and Destination Study, Summary of Findings*, Austin, TX. http://www.austintexas.gov/edims/document.cfm?id=249871. Accessed November 9, 2017.
24. Schaller, B. Enhancing Transit's Competitiveness: A Survey Methodology. *Transportation Research Record: Journal of the Transportation Research Board*, 1999. 1669: 143–149.
25. Morpace International and Cambridge Systematics. *TCRP Report 47: A Handbook for Measuring Customer Satisfaction and Service Quality*. Transportation Research Board of the National Academies, Washington, D.C., 1999.
26. Capital Metro. *Service Plan 2020 - Transit Competitiveness Analysis*, Austin, TX. https://www.capmetro.org/uploadedFiles/Capmetroorg/Future_Plans/Service_Plan_2020/Chapter%202%20-%20Transit%20Competitiveness%20Analysis.pdf. Accessed November 9, 2017.
27. American Planning Association Conference. *Evaluating Transit Competitiveness Using a Transit Competitive Index Tool*, https://www.planning.org/events/event/4199316/. Accessed November 9, 2017.
28. Facts, Figures, and Quotes. *South by Southwest SXSW Conference and Festivals*, Austin, TX. https://www.sxsw.com/facts-figures-quotes/. Accessed October 17, 2017.

The Standing Committee on Transportation Issues in Major Cities (ABE30) peer-reviewed this paper (18-01356).

TRR
JOURNAL OF THE TRANSPORTATION RESEARCH BOARD

Article

Transportation Research Record
2018, Vol. 2672(7) 47–57
© National Academy of Sciences:
Transportation Research Board 2018
Article reuse guidelines:
sagepub.com/journals-permissions
DOI: 10.1177/0361198118792321
journals.sagepub.com/home/trr
SAGE

Bicycle Lanes and Business Success: A San Francisco Examination

Joseph A. Poirier[1]

Abstract

This paper examines the impacts of bicycle infrastructure on business performance in San Francisco, CA. The purpose of the paper is to provide better information about the potential impacts of bicycle infrastructure on different types of business, thereby improving the planning process and providing more confidence in outcomes for stakeholders. The analysis uses secondary data on business employment and sales to measure the change over time in performance of businesses abutting a bicycle lane intervention, versus the performance of nearby but non-abutting businesses. The paper finds that bicycle lane interventions in San Francisco rarely produce profoundly negative outcomes, and that they are often associated with positive business performance outcomes. It also finds that different types of businesses experience bicycle lane interventions differently, with the implication being that transportation planners need to understand and carefully account for the mix of businesses in implementation sites.

As cities across the United States struggle to reduce transportation-related greenhouse gas emissions, they are developing active transportation modes of cycling and walking as key implements within their toolbox of emission-reduction strategies. Cycling holds particular promise, considering approximately 90% of all U.S. trips between 1 and 3 mi are made by automobiles—an eminently bikeable distance (*1*).

The additional societal benefits of cycling include increased physical activity and positive public health outcomes, reduced pollution, lower cost to travelers, and reduced demand on road space, freeing right-of-way for alternative uses (*2–5*).

Research suggests that the availability of bicycle infrastructure (particularly bicycle lanes) is positively associated with bicycle mode share (*6*, *7*). Research has also found that for a given geography, every added bicycle lane mile per square mile (in a typical U.S. city with population over 250,000) is associated with a 1% increase in bicycle commute mode share (*6*, p. 121). Further, research reviewing stated preference surveys found that both people who do and do not bicycle feel safer in and prefer bicycle lanes versus riding with traffic (*8–13*).

The introduction of bicycle lanes in many communities, however, is often met with objection from merchants. Merchants in busy commercial corridors express particular concern with potential loss of business, reasoning that loss of parking spaces, added difficulty loading and unloading freight, and increased congestion caused by bicycle lanes will mean fewer customers and reduced sales (*14–20*).

Pushback from merchants during the planning and implementation processes of bicycle lanes is a significant problem that can delay or cancel projects that would otherwise help address public health, environmental, and traffic congestion issues. This paper uses secondary data from San Francisco to address these concerns and explore associations between bicycle lanes and business success.

Relevance to Practice

Despite the often politically charged planning process that surrounds bicycle infrastructure implementation, very little is known about the actual, quantifiable effects on businesses that can be expected from the introduction of bicycle infrastructure in a commercial corridor. Local planning authorities, therefore, are often put in the difficult position of implementing bicycle infrastructure with little to no contextualizing research. Despite an

[1]Nelson\Nygaard Consulting Associates, Inc., Seattle, WA

Corresponding Author:
Address correspondence to Joseph A. Poirier:
jpoirier@nelsonnygaard.com

abundance of cycling mode choice, safety, and design research conducted in past decades, there has been little quantitative exploration of cycling's effects on commercial establishments.

This lack of research on the economic impacts of bicycle infrastructure, combined with plethoric quantities of research on the positive public health and environmental effects of bicycle infrastructure, leads to intense political pressure to construct bicycle lanes and equally intense political pushback from local commercial interests. Delays caused by this pushback can cost taxpayer dollars, occupy valuable government employee time, and withhold beneficial infrastructure projects from communities. The delays can inject uncertainty into the marketplace, thereby stifling investment or business performance. This research aims to fill these problematic knowledge gaps by investigating how cycling facilities are associated with sales for nearby business establishments—specifically local-serving businesses.

Literature Review

There are few studies of how cycling infrastructure affects economic outcomes for business establishments. This literature review summarizes the existing work, illustrating research needs and shortcomings in planning literature's understanding of the issue. The research that has been conducted suggests that cycling infrastructure is likely positively associated with retail performance but does not draw clear lines of causation between the two.

In perhaps the best-known paper on the economic impacts of bicycle infrastructure, an intercept survey-based examination of various travel mode users in Portland, OR, showed that, on average, cyclists spent more at certain business types and patronized them more often (*21*). This research has been widely celebrated and implemented as a policy tool in bicycle advocacy circles (*22–25*).

Similar survey-based research was carried out in Vancouver, BC, after the installation of separated bikeways, finding that local business sales and profit declined. The paper also concluded that "the moderate negative impact of the lanes will diminish over time as long as mitigation strategies take effect" (*26*, p. iii). The primary difference between the Portland and Vancouver studies is the establishments at which data were collected: in Portland, only high-turnover, sit-down restaurants, 24-h convenience stores, and drinking places were surveyed; in Vancouver, all grade-level establishments on a bicycle lane intervention corridor were surveyed. Although both business operators and customers are prone to reporting error, it is important to consider the difference in conclusions drawn from the two survey methods.

Other survey-based research on shopper mode choice and spending habits has been conducted in the United States. In 2008, a shopper survey in San Francisco's Columbus Avenue corridor found that those who bicycle, ride in taxis, or use "other" modes of transportation, spent more in the corridor than those who drove or took transit (*27*, pp. 5–7). In New York City's East Village, a 2012 survey study concluded that total aggregate spending of bicyclists in the neighborhood was more than drivers', bus riders', taxi users', and "select bus service" riders' (*28*, p. 9). The same survey also found that cyclists had the highest per capita spending in the neighborhood (*28*, p. 10). This high level of cyclist spending was attributed partially to the fact that cyclists visited local businesses more often and also to the addition of protected bicycle lanes on 1st and 2nd Avenues (*28*, pp. 14–15).

In a 2008 survey-based examination of cycling- and car-borne customers at inner Melbourne, Australia, shopping strips, Lee argued that if public auto parking spaces in commercial areas were reallocated to cyclist parking, 3.6 times the current retail spend could be achieved in the area, and that this would be especially advantageous for cafés, restaurants, and clothing retailers. Lee's argument was based on a customer survey conducted in Melbourne which estimated auto and cyclist spending behavior. Lee then multiplied the average reported cyclist spend by the number of cyclists that could park in a single auto parking space (*29*, p. 39). Next, the author compared the multiplied cyclist spend product to the average spend of an auto driver occupying the same amount of space, producing the 3.6 × value of cyclists (*29*, p. 39). This line of reasoning, although perfectly logical, is fraught with assumption and context-specific dependencies. For example, Lee assumes that there is an unlimited number of cyclists waiting in the wings to take the place of auto users and also that the mixture of local businesses would remain unchanged with the loss of auto parking. Further, the paper's geographic context is highly specific—"inner suburban Melbourne" has a particular built environment, climate, and demographic mix, making it difficult to generalize results to other communities (a problem with any small geography study of this kind) (*29*, p. 8).

Dual survey-based studies performed on the Bloor Street corridor in Toronto found that, overall, merchants and shoppers alike supported the addition of bicycle lanes to the right-of-way and that those arriving by bicycle, transit, and foot were likely to spend more in the commercial district than those arriving by auto (*30*, *31*). Survey results from the 2009 research revealed that "75% of merchants thought a bike lane or widened sidewalk would improve or have no effect on business, and patrons preferred a bike lane to widened sidewalks at a ratio of almost four to one" (*30*, p. 23). The 2009 analysis roundly concluded "that merchants in this area are unlikely to be negatively affected by reallocating on-street parking space to a bicycle lane. On the contrary, this change will likely increase commercial activity" (*30*, p. 1).

In 2010, a similar survey-based analysis conducted in the area found that more than half of merchants believed reducing parking and adding a bike lane would "increase or have no impact on their daily number of customers" (*31*, p. 6). The same study found that "The majority of people surveyed (58%) preferred to see street use reallocated for widened sidewalks or a bike lane, even if on-street parking were reduced by 50%." These two analyses—as with most of the research cited in this literature review—measure only perception of merchants and customers, not hard sales data or productivity measures.

Other, non-survey-based, research projects have tackled the question of cycling's economic impact. Cortright, in 2007, produced a white paper arguing that Portland, OR, residents save $1.5 billion annually because of their shorter average commute distances (*32*). Cortright posits that because transportation expenses have a much lower local multiplier effect than money not spent on transportation, Portland's investment in lower-cost transport modes (such as cycling) produces a greater residual multiplier effect in the local economy than, say, single-occupancy vehicle driving (*32*). This paper represents a novel argument in cycling economic impact research and deserves to be considered more frequently.

A 2003 interview-based study examined bicycle lanes added to San Francisco's Valencia Street, uncovering strong merchant support for the lanes, including the belief among 65% of interviewees that bicycle lanes have had positive impacts on sales and business overall (*33*). It is important to note that industrial and office business types were not included in the interviews. This finding will be returned to in this paper's discussion section, as Valencia is a case study in this research.

Some research on the economic impacts of cycling uses sales data. A 2012 New York City (NYC) Department of Transportation (DOT) publication claimed increases in retail sales of "up to 49%" along new protected bicycle lanes on 9th Avenue from 23rd to 31st streets, compared with increases of 3% borough-wide (*34*, p. 4). Similarly, a 2017 Oakland DOT document using sales tax receipt data claimed retail sales in the Telegraph corridor had increased 9% year over year after a protected bicycle lane intervention (*35*). However, in neither DOT study were the bicycle lane corridors compared with control corridors or other nearby small-geography areas.

A taxable retail sales-based analysis of two bicycle lane interventions in Seattle included dual control regions for each of the two study areas, showing no negative sales impact on businesses resulting from either intervention (*36*). The study suggested the possibility of a wildly successful economic impact produced by a climbing lane that removed 12 parking spots and was correlated with a 400% increase in sales in the district (*36*, p. 2). The Seattle research used a more sound and transparent methodology than both the NYC and Oakland DOT reports, with more appropriate control regions and greater discussion of the results (*34*, *35*).

Other academic studies and news reports have documented business dissatisfaction with bicycle infrastructure, focusing on the fact that parking spaces are often removed or freight access is limited when bicycle lanes are implemented (*14*, *15*, *37*). However, much of the evidence in these papers is anecdotally-based and relies on perceptions of change, demonstrating a clear need for more sales and business performance-based research.

A number of 'grey literature' (non-traditional, often commercially- or advocacy-produced papers) items have recirculated the above-reviewed original research on the relationship between bicycle infrastructure and nearby business outcomes, thereby artificially inflating the number of 'sources' of information that bicycle advocates, city planners, and other pro-bicycle infrastructure actors are able to wield in attempts to convince business owners that bicycle lanes increase sales (*25*, *38–42*). Extra-economic effects of added bicycle infrastructure may alone justify such infrastructure's construction, but the extent to which reports and papers have recycled and generalized the findings of a few corridor-level research endeavors is viewed as somewhat disingenuous by this author.

Although there is a slew of research showing the economic impacts of tourism cycling (e.g. *43–49*), almost none of it helps policy-makers understand what the expected impacts of bicycle lane projects are for small businesspeople in an urban context. This is a clear and necessary research need and one that this project readily addresses, albeit in a San Francisco context.

Methods

This analysis uses longitudinal firm-level data from the National Establishment Time-Series (NETS) data set, constructed from Dun & Bradstreet data, to track establishment sales over time on three corridors in which bicycle lanes were recently implemented in San Francisco: Columbus Avenue, Polk Street, and Valencia Street (*50*).

These three bicycle lane intervention case study sites were selected based on data availability, nearby business establishment density, date of bicycle lane implementation, and bicycle lane intervention type.

To be eligible for selection, corridors must have experienced a bike lane intervention (defined as a major redesign of a right-of-way that significantly improves the ease of cycling) within a time period covered by the NETS database. This temporal criteria required NETS data to be available for 2 years before and after the bike lane intervention. The pool of available cases was therefore limited because the NETS data available to me covered 1990 to 2013 (ruling out bicycle lanes implemented before 1992 or after 2011). Many of the more recent Bay Area bike lane interventions, such as the Telegraph Avenue protected bike lanes in Oakland and Puerto Suello Hill Path in San Rafael, were ineligible for selection because of the lack of NETS data.

I used online news articles, blog postings, and Google Earth Pro to identify bicycle lane intervention sites that satisfied the above criteria. Google Earth Pro's historical satellite imagery was especially helpful because bicycle lane changes at intervals as precise as 1 month were discernable.

After intervention sites that met the suitable temporal and infrastructure implementation requirements were selected, basic descriptive statistics were executed to identify sites with a suitable number of establishments for analysis. The Berkeley Bicycle Boulevard program, which was largely implemented in 2002–2003, was ruled out because there were no business establishments abutting the bicycle boulevards that existed for 2 years before and 2 years after the intervention (*51*). Oakland's 2nd Street Class III bicycle lane in Jack London Square, which was painted in 2011, was similarly ruled out.

Final Study Areas

Three study areas were thus identified for the analysis: Valencia Street, Columbus Avenue, and Polk Street corridors in San Francisco. All three corridors received a bicycle lane intervention and had a significant number of business establishments present for 2 years before and after the intervention. Descriptions of each study area follow and include the number of establishments that abut the bicycle lane intervention and the number of establishments that are nearby (within adjacent block groups) but are not abutting the bicycle lane intervention.

Study Area 1: Valencia Street. Valencia Street is a bustling commercial corridor in the Mission District. The 1999 intervention here turned a road with two traffic lanes in either direction into a thoroughfare with one traffic lane in either direction, one bicycle lane in either direction, and one turning lane in the middle of the street (*33*, p. 4). This study area is of particular value in evaluating this paper's methodology, as an interview-based economic impact analysis (*33*) and a 1-year traffic impact analysis (*52*) have been conducted for this same bicycle lane intervention. There were 224 establishments abutting the bicycle lane intervention from 1997 to 2001, and 955 non-abutting establishments. Figure 1 shows the Valencia, Polk, and Columbus study areas.

Study Area 2: Columbus Avenue. Columbus Avenue is a major four-lane thoroughfare from the Fisherman's Wharf neighborhood to the Financial District. Both of these neighborhoods are popular with tourists and are densely populated with residences and business establishments. There are parking lanes on both sides of the roadway.

In 2005, Columbus Avenue was re-striped and painted to include sharrows on the roadway. The intervention stretched from North Point Street in the Fisherman's Wharf neighborhood, south to Washington Street in the Financial District. There were 143 permanent establishments abutting this stretch of Columbus Avenue from 2003 to 2007, and 1,225 non-abutting establishments.

Study Area 3: Polk Street. Polk Street is a two-lane commercial north–south corridor connecting Ghirardelli Square with Market Street. There is one parking lane on either side of the roadway. Using Google Earth Pro, I found that sharrows and painted bicycle lanes were added in 2004 to the right-of-way between Post and Union Streets, transforming the street into a Class II/III bicycleway. There were 212 permanent establishments on this stretch of Polk Street from 2002 to 2006, and 645 non-abutting establishments. Figure 1, below, shows the Valencia, Columbus, and Polk study areas.

Establishment Selection

Next, I analyzed change over time of nearby business establishment performance at the block group level in the three selected case study areas using two major categories of establishment: all establishments and local-serving establishments. For each establishment type, the change over time analysis was conducted for "abutting" and "non-abutting" addresses.

Abutting establishments, or establishments facing the street that saw the intervention, were identified by address in the NETS database. All establishments with the intervention street's name in their mailing address were flagged with a dummy indicator as abutting the target street (e.g., all establishment mailing addresses that included the word "Valencia" were flagged as abutting the street that saw the intervention). All other establishments (those not flagged as abutting) were denoted using the same dummy variable as non-abutters.

To more appropriately explore if establishment types played a meaningful intervening role in the analysis, I subclassified the establishments, using North American Industry Classification System (NAICS) codes, into a

Figure 1. Valencia Street study area and establishments.

"local-serving" group. The methodology used to classify the establishments into these two categories is derived from two previous studies and is modified to include business establishments that might reasonably be patronized by cyclists (i.e. furniture stores were not included, as cyclists are unlikely to purchase couches during bicycle trips) (*53*, *54*). Previous studies adapted their method of restricting research to subsets of NAICS categories from still earlier work (*55*, *56*) and individually selected those NAICS deemed local-serving in nature (*53*, pp. 6–7). Some past researchers identified four industry categories they believed were "local-serving" in nature and also used two-digit NAICS prefixes to isolate retail and food service and hospitality (*54*, p. 76). I found both of these definitions of local-serving too limiting and not inclusive enough of businesses that could reasonably serve nearby customers via bicycle. Footwear and leather goods stores, for example, were not included in either of the primary research precedents, despite the ease with which a shoebox can be carried on a bicycle and the local-serving

Table 1. Local-Serving NAICS-Based Classification

NAICS code	Description	NAICS code	Description
443142	Electronics stores	4512-	Book stores and news dealers
44413-	Hardware stores		Book stores
445-	Food and beverage stores		News dealers and newsstands
	Supermarkets and other grocery (except convenience) stores	4521-	Department stores
	Convenience stores	4531-	Florists
	Meat markets	4532-	Office supplies, stationery, and gift stores
	Fish and seafood markets		Office supplies and stationery stores
	Fruit and vegetable markets		Gift, novelty, and souvenir stores
	Baked goods stores	4533-	Used merchandise stores
	Confectionary and nut stores	45391-	Pet and pet supply stores
	All other specialty food stores	453991	Tobacco stores
	Beer, wine, and liquor stores	453998	All other miscellaneous store retailers
446-	Health and personal care stores	522120	Savings institutions
	Pharmacies and drug stores	52213-	Credit unions
	Cosmetics, beauty supplies, and perfume stores	52219-	Other depository credit intermediation
	Optical goods stores	522291	Consumer lending
	Food (health) supplement stores	722330	Mobile food services
	All other health and personal care stores	7224-	Drinking places (alcoholic beverages)
448-	Clothing and clothing accessories stores	7225-	Restaurant and other eating places
	Men's clothing stores		Full-service restaurants
	Women's clothing stores		Limited-service restaurants
	Children and infants' clothing stores		Cafeterias, grill buffets, and buffets
	Family clothing stores		Snack and nonalcoholic beverage bars
	Clothing accessories stores	811213	Communication equipment repair and maintenance
	Other clothing stores	81143-	Footwear and leather goods repair
	Shoe stores	81211-	Hair, nail, and skin care services
	Jewelry stores		Beauty salons
	Luggage and leather goods stores		Nail salons
4511-	Sporting goods, hobby, and musical instrument stores	81231-	Coin-operated laundries and drycleaners
	Sporting goods stores	81232-	Drycleaning and laundry services (except coin-operated)
	Hobby, toy, and game stores	81292-	Photofinishing
	Sewing, needlework, and piece goods stores		

nature of the business (*53*, *54*). NAICS codes included in my analysis are shown in Table 1.

Change Over Time Analysis

After identifying study areas and relevant establishment types, two outcome metrics were identified for each category of establishment: sales per establishment and sales per employee. Although sales per establishment may solely represent the size or sales volume of the establishment and have little bearing on the establishment's success, it is included as a general measure and one with which sales per employee (a better productivity measure) can be compared and contrasted.

First, all dollar figures were Consumer Price Index for all Urban Consumers (CPI-U) adjusted to 2013 U.S. dollars. Second, for each study area block group, establishment density, average employee per establishment counts, average sales per establishment counts, and average sales per employee counts were calculated for all abutting and non-abutting establishments and abutting and non-abutting local-serving establishments.

In the afore- and after-mentioned averages, establishments with one or fewer employees and/or annual revenue of more than $50,000,000 were considered outliers (either small, household-type incorporations, or very large companies) and excluded from annual calculations. This follows in the footsteps of other responsible uses of NETS data (*57*, p. 56). Table 2 details dropped establishments.

Further, only establishments present for the complete 5-year period were included, thereby ensuring that the measurements captured change in performance for businesses that existed before the bike lane intervention, and were not capturing the effects of new businesses moving into the study area. Drawbacks to this method are explored in the discussion section.

After isolating the required business establishments, I mean averaged their change in outcome metrics across all

Table 2. Establishments Dropped from Analysis

	Valencia	Columbus	Polk
Establishments with ≤ 1 employee	297	383	219
Establishments dropped (%)	20	22	20
Establishments with > $50,000,000 annual revenue	11	15	3
Establishments dropped (%)	1	1	0
Totals			
Total establishments	1,487	1,765	1,078
Total establishments dropped	308	397	221
Establishments dropped (%)	21	22	21

Table 3. All Abutting Establishment Performance over 5-Year Period versus All Non-Abutting Establishments

	Sales per establishment	Sales per employee
Valencia Street	+ 1.73%	+ 0.94%
Columbus Avenue	−15.11%	+ 0.09%
Polk Street	−5.77%	−1.02%

Table 4. Abutting Local-Serving Establishment Performance over 5-Year Period versus Non-Abutting Local-Serving Establishments

	Sales per local-serving establishment	Sales per local-serving employee
Valencia Street	+ 38.33%	+ 3.44%
Columbus Avenue	+ 5.86	−4.08%
Polk Street	+ 21.56%	+ 4.33%

of the study area block groups, using a frequency weighting of number of establishments.

After computing change over time statistics for the three study areas and their establishment types, change over time in the abutting establishments was compared with change over time for the non-abutting establishments in the same study area. These numbers were developed as absolute dollar differences and as percentage change differences. The difference in percentage change figures are more meaningful measures as they account for the variation in average establishment sales at the beginning of the study period, whereas absolute dollar difference does not.

Discussion

The results and implications of this exercise are inherently limited by the geographic, cultural, climatic, and other peculiarities of the study areas in which the analysis was conducted. In short, every transportation corridor is different and there can be no expected uniform result of a bicycle lane intervention. This analysis can, however, supply evidence to decision makers looking to compare upcoming with past projects and estimate effects of future interventions.

The analysis is separated into "all" and "local-serving" establishment categories and shows a few distinct associations of bicycle lane interventions with nearby business establishment sales. For one, the "all establishment" results (Table 3) show varied relationships on business performance. The low positive figures for Valencia and Columbus sales per employee are important because they mean that a small amount of error could flip the results directionally. These generally low-magnitude findings suggest that it is unlikely that a blanket "bicycle lanes are good for all business" statement is accurate.

Local-serving establishments, however, show a largely positive relationship between bicycle lane intervention and establishment performance (Table 4). Sales per establishment, in particular, operate strongly in favor of a suggestion that bicycle lane interventions are good for local-serving businesses. Sales per employee do not show nearly as strong an advantage for local-serving abutting establishments over local-serving non-abutting establishments.

The results of this study corroborate some findings in previous economic impact studies of bicycle lane interventions. The 2003 interview-based study of the Valencia bicycle lane intervention, for example, showed that 65% of nearby merchants believed the bicycle lane intervention had a positive effect on business (*33*, p. 7). Tables 3 and 4, above, show that this is also borne out in NETS data for abutting establishments, strengthening the argument that this intervention resulted in positive business outcomes.

The notion that some bicycle lane interventions may have had negative impacts on abutting establishments is supported somewhat by the "all establishments" metrics in Table 3, in which Polk figures are negative and Valencia figures only slightly positive. All establishments abutting the Columbus intervention also showed a considerable relative decline in sales per establishment, as did local-serving establishments' sales per employee. The decline in sales per establishment for all establishments in the Columbus study area is the most extreme negative result in this research.

These results match with mixed results of past survey-based economic impact analyses of bicycle lane interventions, whereby some research has shown positive (or likely positive) effects of interventions (*21, 30, 31, 33–35*) and one showed negative effects (*26*). It should be noted that the negative effects of bicycle lane interventions suggested by this research are largely not extreme.

The advantage of the analysis carried out in this paper is that NETS data provide a more objective look at the relationship between bicycle lane interventions and business success than, say, survey- or interview-based analyses, which measure perception, not real effect. The use of appropriate control regions (neighborhood non-abutting establishments) is a significant methodological improvement over the two DOT publications reviewed earlier (*34, 35*).

Additional research is needed to determine whether the taxable sales index used by Rowe is a more meaningful measure than NETS (*36*). That being said, the two most robust sales-based analyses of bicycle lane interventions (Rowe [*36*] and this analysis) have strongly suggested that bicycle lane interventions do not have profound negative impacts and may have significant positive impacts.

Methodological Shortcomings

There are several shortcomings with this methodology. For one, bicycle lane interventions are not arbitrary actions undertaken by municipal bureaucracies. They are often contextualized within a changing social, cultural, or political landscape, meaning that assigning changes in productivity or revenues to bicycle lane interventions is nearly impossible using this type of analysis. A business district seeing improved foot traffic or facing gentrification pressures may lobby for the addition of a bicycle lane, or a bicycle lane may arrive at a similar time as other neighborhood demographic changes. In short, there is a self-selection problem here. By comparing abutting establishments with non-abutting establishments within the study area, much of this greater neighborhood change can be accounted for—but not all.

Another of this analysis' flaws is the absence of parking data. A chief complaint of merchants abutting bicycle lanes is that the reduction in parking will reduce their customer base (*14, 16–20*). This paper does very little to research this concern, except to proxy a bicycle lane intervention for increased difficulty of customer parking. Even this assumption is tenuous.

Using addresses to locate establishments also causes issues, as mailing addresses and storefront entrances are not always intuitively correlated. Corner establishments, in particular, may be over- or underrepresented in the abutting establishments category. For example, establishments located at the corner of Valencia and 19th—although certainly abutting a bicycle lane—may not be listed on Valencia Street in the NETS data.

NETS employment data should be transformed to 3-year averages for more robust analysis (*58*, p. 15). This would have been a preferred method for the case study regions at hand but was infeasible because of the already small number of establishments produced by the 2-year before-and-after constraint. I deemed it appropriate to trade the larger sample size for a slightly increased margin of error, particularly given the 5-year study period.

By analyzing only establishments that existed for the 5-year study period, I introduced survival bias and probably overrepresented businesses that were predisposed to succeed under the conditions of bicycle lanes. Businesses that suffered great revenue losses or found the freight loading situation cripplingly difficult, for example, could have moved or gone out of business. This analysis captures neither of these phenomena. Similarly, the analysis may underreport businesses that opened or moved in because of the presence of bicycle lanes. If two bicycle shops moved to the study area during the intervention year for proximity to target customers, for example, that outcome would not have been captured.

Finally, this paper makes the major assumption that businesses directly abutting the bicycle lane will see the greatest benefit, whereas businesses a block away from the bicycle lane will not. Although there is logic behind this assumption, it is easy to imagine a cycling trip that continues an additional 50 to 100 Ft after leaving a bicycle lane with little impedance borne by the traveler.

Conclusions

- By only one metric (sales per establishment), in one study area, did the bicycle lane intervention have a profound negative impact. Bicycle lanes do not seem to have the catastrophic negative effect on business that some merchants claim.
- For the Valencia and Polk bicycle lane interventions, sales per local-serving establishment increased markedly after the intervention, suggesting bicycle lanes can have a positive effect on local-serving businesses.
- If these three cases were to be representative, it is not true to claim that bicycle lanes will increase sales for all businesses. Put more simply, a blanket

"bicycle lanes are good for all businesses" statement is likely untrue.
- Local-serving and all business establishments probably experience bicycle lane interventions differently. Care should be taken by planners to consider who should benefit from bicycle lane interventions. This research suggests that local-serving businesses benefit more than all businesses.

Future Research Recommendations

Although this method used neighborhood-level geographies as control areas, its findings would be strengthened with additional controls—notably, adjacent corridors that did not experience bike lane interventions. Future research should also narrow the scope to address Class I, II, or III lanes individually, thereby producing results with more specific actionable results. Lastly, parking is an important piece of this study and was not included in this research. Future work should include parking data to help explain the economic impacts of parking loss that occur with bicycle lane interventions.

Author Contributions

The author confirms sole contribution to the paper as follows: study conception and design; data collection; analysis and interpretation of results; and draft manuscript preparation. The author reviewed the results and approved the final version of the manuscript.

References

1. Carr, L. J., S. I. Dunsiger, and B. H. Marcus. Validation of Walk Score for Estimating Access to Walkable Amenities. *British Journal of Sports Medicine*, Vol. 45, No. 14, 2010, pp. 1144–1148. doi:10.1136/bjsm.2009.069609.
2. Fishman, E., P. Schepers, and C. B. M. Kamphuis. 2015. Dutch Cycling: Quantifying the Health and Related Economic Benefits. *American Journal of Public Health*, Vol. 105, No. 8, 2015, pp. e13–e15. doi:10.2105/AJPH.2015.302724.
3. Götschi, T., J. Garrard, and B. Giles-Corti. Cycling as a Part of Daily Life: A Review of Health Perspectives. *Transport Reviews*, Vol. 36, No. 1, 2016, pp. 45–71. doi:10.1080/01441647.2015.1057877.
4. Pucher, J., R. Buehler, D. R. Bassett, and A. L. Dannenberg. Walking and Cycling to Health: A Comparative Analysis of City, State, and International Data. *American Journal of Public Health*, Vol. 100, 2010, pp. 1986–1992. doi:10.2105/AJPH.2009.189324.
5. Audrey De, N., M. J. Nieuwenhuijsen, J. M. Antó, M. Brauer, D. Briggs, C. Braun-Fahrlander, N. Cavill, A. R. Cooper, H. Desqueyroux, S. Fruin, G. Hoek, L. I. Panis, N. Janssen, M. Jerrett, M. Joffe, Z. J. Andersen, E. van Kempen, S. Kingham, N. Kubesch, K. M. Leyden, J. D. Marshall, J. Matamala, G. Mellios, M. Mendez, H. Nassif, D. Ogilvie, R. Peiró, K. Pérez, A. Rabl, M. Ragettli, D. Rodríguez, D. Rojas, P. Ruiz, J. F. Sallis, J. Terwoert, J. -F. Toussaint, J. Tuomisto, M. Zuurbier, and E. Lebret. Improving Health through Policies That Promote Active Travel: A Review of Evidence to Support Integrated Health Impact Assessment. *Environment International*, Vol. 37, No. 4, 2011, pp. 766–77. doi:10.1016/j.envint.2011.02.003.
6. Dill, J., and T. Carr. Bicycle Commuting and Facilities in Major U.S. Cities: If You Build Them, Commuters Will Use Them. *Transportation Research Record: Journal of the Transportation Research Board*, 2003. 1828: 116–123.
7. Buehler, R., and J. Pucher. Cycling to Work in 90 Large American Cities: New Evidence on the Role of Bike Paths and Lanes. *Transportation* Vol. 39, 2012, pp. 409–432.
8. Akar, G., and K. Clifton. Influence on Individual Perceptions and Bicycle Infrastructure on Decision to Bike. *Transportation Research Record: Journal of the Transportation Research Board*, 2009. 2140: 165–172.
9. Antonakos, C. Environmental and Travel Preferences of Cyclist. *Transportation Research Record: Journal of the Transportation Research Board*, 1994. 1438: 25–33.
10. Fishman, E., S. Washington, N. Haworth, and A. Watson. Factors Influencing Bike Share Membership: An Analysis of Melbourne and Brisbane. *Transportation Research Part A*, Vol. 71, 2015, pp. 17–30.
11. Landis, B., V. Vattikuti, and M. Brannick. Real-Time Human Perceptions: Toward a Bicycle Level of Service. *Transportation Research Record: Journal of the Transportation Research Board*, 1997. 1578: 119–126.
12. Sanders, R. L., and J. F. Cooper. Do All Roadway Users Want the Same Things? Results from Roadway Design Survey of San Francisco Bay Area Pedestrians, Drivers, Bicyclists, and Transit Users. *Transportation Research Record: Journal of the Transportation Research Board*, 2013. 2393: 155–163.
13. Sener, I. N., N. Eluru, and C. R. Bhat. An Analysis of Bicycle Route Choice Preferences in Texas, US. *Transportation*, Vol. 36, No. 5, 2009, pp. 511–539. doi:10.1007/s11116-009-9201-4.
14. Boone, A. San Jose Merchants Object to Parking Removal for Bike Lanes on Park Ave | Streetsblog San Francisco. *StreetsBlog SF*, August 19, 2014. http://sf.streetsblog.org/2014/08/19/san-jose-merchants-object-to-parking-removal-for-bike-lanes-on-park-ave/.
15. Curry, M. Oakland's New Parking Protected Bike Lanes Are Challenging to Some | Streetsblog San Francisco. *StreetsBlog SF*, May 4, 2016. http://sf.streetsblog.org/2016/05/04/oaklands-new-parking-protected-bike-lanes-are-challenging-to-some/.
16. Mirk, S. Businesses Protest Planned Downtown Bike Lane. *Portland Mercury*, June 6, 2012. http://blogtown.portlandmercury.com/BlogtownPDX/archives/2012/06/06/businesses-protest-planned-downtown-bike-lane.
17. Jaffe, E. Here We Go Again With Retailers Hating Bike Lanes. *Citylab*, October 7, 2015. http://www.citylab.com/commute/2015/10/here-we-go-again-with-retailers-hating-bike-lanes/409372/.
18. Peterson, H. Merchants Protest New Downtown Bike Lanes. *Washington Examiner*, October 27, 2010. http://

www.washingtonexaminer.com/merchants-protest-new-downtown-bike-lanes/ article/108847.
19. Bialick, A. SFMTA Cuts Block of Polk Bike Lane Fought By Visionless Mayor's Optometrist. *StreetsBlog SF*, February 26, 2015. http://sf.streetsblog.org/2015/02/26/sfmta-cuts-block-of-polk-bike-lane-fought-by-visionless-mayors-optometrist/.
20. Del Signore, J. Two Anti-Bike Lane Protests Aim to Eradicate Bike Lanes. *The Gothamist*, October 24, 2010. http://gothamist.com/2010/10/14/two_anti-bike_lane_protests_aim_to.php.
21. Clifton, K., K. Currans, C. Muhs, C. Ritter, S. Morrissey, and C. Roughton. Consumer Behavior and Travel Choices: A Focus on Cyclists and Pedestrians. National Association of City Transportation 2012. https://nacto.org/docs/usdg/consumer_behavior_and_travel_choices_clifton.pdf.
22. Maus, J. Study Shows Biking Customers Spend More. *BikePortland.org*, July 6, 2010. https://bikeportland.org/2012/07/06/study-shows-biking-customers-spend-more-74357.
23. Campbell, D. People Mean Business. *Bike East Bay*, July 25, 2015. https://bikeeastbay.org/people-mean-business.
24. Max. Bicycles Are Business: What Research Says About Bicycling's Economic Benefits. *Bicycle Coalition Blog*, July 23, 2014. http://bicyclecoalition.org/facts-biking-improves-business/#sthash.Q66Bjqpo. dpbs.
25. Szczepanski, C. How Bicycles Bring Business. *Momentum Mag*, April 10, 2013. https:// momentummag.com/how-bicycles-bring-business/.
26. Stantec Consulting Ltd. *Vancouver Separated Bike Lane Business Impact Study*. Stantec Consulting Ltd, Vancouver, BC, 2011.
27. San Francisco County Transportation Authority. *Columbus Avenue Neighborhood Transportation Study*. San Francisco County Transportation Authority, Francisco, San CA, 2010. http://www.sfcta.org/sites/default/files/content/Planning/ColumbusAvenue/FinalReport/SFCTA%20Columbus%20Ave%20Final%20Report_lo_res.pdf.
28. Transportation Alternatives. *East Village Shoppers Study: A Snapshot of Travel and Spending Patterns of Residents and Visitors in the East Village*. Transportation Alternatives, New York City, NY, 2012.
29. Lee, A. *What is the Economic Contribution of Cyclists Compared to Car Drivers in Inner Suburban Melbourne's Shopping Strips?* Masters of Urban Planning thesis, The University of Melbourne, Melbourne, Australia, 2008.
30. Sztabinski, F. *Bike Lanes, On-Street Parking and Business: A Study of Bloor Street in Toronto's Annex Neighbourhood*. Clean Air Partnership, Toronto, 2009.
31. Forkes, J., and N. S. Lea. Bike Lanes, On-Street Parking & Business: Year 2 Report: A Study of Bloor Street in Toronto's Bloor West Village.Clean Air Partnership, Toronto, 2010.
32. Cortright, J. *Portland's Green Dividend*. CEOs for Cities, 2007.
33. Drennan, E. *Economic Effects of Traffic Calming on Urban Small Businesses*. San Francisco, CA, 2003.
34. New York City Department of Transportation. *Measuring the Streets: New Metrics for 21st Century Streets*. Department of Transportation, City of New York, NY, 2012.
35. Fine, S., and N. Cartagena. *Telegraph Avenue Progress Report*. Progress Report. City of Oakland Department of Transportation, Oakland, CA, 2017. http://www2.oaklandnet.com/oakca1/groups/pwa/ documents/report/oak062598.pdf.
36. Rowe, K. *Bikenomics: Measuring the Economic Impact of Bicycle Facilities on Neighborhood Business Districts*. University of Washington, Seattle, Washington, 2013.
37. Pagliara, F., and L. Biggiero. *A Survey on Impacts of Bike-Oriented Policies on Residents and Economic Activities*. WIT Transactions on Ecology and the Environment 191, 2014. doi:10.2495/SC140631.
38. Tolley, R. *Good for Busine$$: The Benefits of Making Streets More Walking and Cycling Friendly*. National Heart Foundation of Australia, 2011.
39. Flusche, D. *The Economic Benefits of Bicycle Infrastructure Investments*. League of American Bicyclists, 2009. http://www.advocacyadvance.org/docs/economic_benefits_bicycle_infrastructure_report. pdf.
40. Andersen, M., and M. L. Hall. *Protected Bike Lanes Mean Business*. People for Bikes, Alliance for Biking & Walking, 2014. http://www.bikewalkalliance.org/resources/reports/protected-bike-lanes-mean-business.
41. Flusche, D. *Bicycling Means Business: The Economic Benefits of Bicycle Infrastructure*. League of American Bicyclists & Alliance for Biking & Walking, 2012.
42. Arancibia, D. *Cycling Economics: Economic Impacts of Bike Lanes*. Toronto, Canada. http://www.torontocycling.org/uploads/1/3/1/3/13138411/cycling_economies_eglinton_final.pdf. Accessed April 10, 2017.
43. Cook, T. J., S. W. O'Brien, K. N. Jackson, S. Searcy, and D. J. Findley. *Behavioral Effects of Completing a Critical Link in the American Tobacco Trail*. Institute for Transportation Research and Education, North Carolina State University, Raleigh, NC, 2014. https://itre.ncsu.edu/wp-content/uploads/2016/03/American-Tobacco-Trail-Final-Report-ITR-2014.pdf.
44. Meletiou, M., J. Lawrie, T. Cook, S. O'Brien, and J. Guenther. Economic Impact of Investments in Bicycle Facilities: Case Study of North Carolina's Northern Outer Banks. *Transportation Research Record: Journal of the Transportation Research Board*, 2005. 1939: 15–21.
45. Grabow, M., M. Whited, and M. Hahn. *Valuing Bicycling's Economic and Health Impacts in Wisconsin*. Nelson Institute for Environmental Studies University of Wisconsin School of Medicine and Public Health, November 4, 2010. ftp://ftp2.dot.state.mn.us/pub/outbound/transit/NonMotorized%20Transportation%20Research%20Committee/Grabow_11.04.10presentation.pdf.
46. Bowker, J., J. Bergstrom, J. Gill, and U. Lemanski. The Washington & Old Dominion Trail: An Assessment of User Demographics, Preferences, and Economics. *Headwaters Economics*, 2004. https://headwaterseconomics.org/trail/21-washington-and-old-dominion-trail/.
47. East Central Florida Regional Planning Council. *Economic Impact Analysis of Orange County Trails*. East Central Florida Regional Planning Council, 2011. https://www.dep.state.fl.us/gwt/economic/PDF/Orange_County_Trail_Report_final_May2011.pdf.

48. Bowker, J. M., J. C. Bergstrom, and J. Gill. Estimating the Economic Value and Impacts of Recreational Trails: A Case Study of the Virginia Creeper Rail Trail. *Tourism Economics*, Vol. 13, No. 2, 2007, pp. 241–260. doi: 10.5367/000000007780823203.
49. Kaliszewski, N. *Jackson Hole Trails Project Economic Impact Study*. University of Wyoming, Laramie, Wyoming, 2011. https://headwaterseconomics.org/wp-content/uploads/Trail_Study_16-jackson-hole-trail-project.pdf.
50. National Establishment Time-Series (NETS) Database. *Kauffman Compendium on Entrepreneurship and Innovation Data*, September 2016. http://www.kauffman.org/~/media/kauffman_org/microsites/sotf/data_sources/NETS_data_overview.pdf.
51. City of Berkeley. Bicycle Boulevard History - City of Berkeley, CA, 2016. https://www.cityofberkeley.info/Public_Works/Transportation/Bicycle_Boulevard_History.aspx. Accessed May 4, 2016.
52. Sallaberry, M. *Valencia Street Bike Lanes: A One Year Evaluation*. The Department of Parking & Traffic, San Francisco, CA, 2000. http://industrializedcyclist.com/Valencia_bikelanes.pdf.
53. Plowman, J. *What Happens to Retail When Neighborhoods Gentrify? Evidence from California Metro Areas*. University of California, Berkeley, Professional Report, Berkeley, CA, 2014.
54. Meltzer, R., and J. Schuetz. Bodegas or Bagel Shops? Neighborhood Differences in Retail and Household Services. *Economic Development Quarterly*, Vol. 26, No. 1, 2012, pp. 73–94. doi:10.1177/0891242411430328.
55. Koebel, C. T. *Analyzing Neighborhood Retail and Service Change in Six Cities*. Technical Report. Virginia Center for Housing Research, 2014. https://vtechworks.lib.vt.edu/handle/10919/48643.
56. Chapple, K. *Analysis of Commercial Trends, Mission and 24th Streets*. Unpublished Report. Center for Community Innovation, UC Berkeley, 2009.
57. Chapple, K., and R. Jacobus. Retail Trade as a Route to Neighborhood Revitalization. In *Urban and Regional Policy and Its Effects*, II. Brookings Institutions Press, Washington, D.C, 2009, pp. 19–68.
58. Neumark, D., J. Zhang, and B. Wall. *Business Establishment Dynamics and Employment Growth*. 05-02. Entrepreneurship Seminar Series. Hudson Institute Center for Employment Policy, 2005. http://ssrn.com/abstract=909249.

The Standing Committee on Transportation Issues in Major Cities (ABE30) peer-reviewed this paper (18-03763).

Article

TRR
JOURNAL OF THE TRANSPORTATION RESEARCH BOARD

Transportation Research Record
2018, Vol. 2672(7) 58–68
© National Academy of Sciences:
Transportation Research Board 2018
Article reuse guidelines:
sagepub.com/journals-permissions
DOI: 10.1177/0361198118797215
journals.sagepub.com/home/trr
SAGE

Sensors and the City: Urban Challenges for Parking Occupancy Detection and Pricing

Soumya S. Dey[1], Stephanie Dock[1], Alek Pochowski[2], Meredyth Sanders[2], Benito O. Pérez[1], Matt Darst[3], and Eduardo Cardenas Sanchez[4]

Abstract

Implementing performance parking using demand-driven hourly parking meter rates and real-time occupancy information improves the customer experience and provides more available on-street parking in selected cities across the United States. The implementation of performance parking is not necessarily simple, however, and planners must overcome a host of challenges posed by the urban environment. This paper discusses an approach with the potential to become the state-of-the-practice for developing real-time availability for on-street parking in an urban environment, including how the District Department of Transportation in Washington, D.C., dealt with challenges related to: sensor communication, on-street activity, special events, limited space, broader mobility issues, parking users, coordinating installation, and interagency coordination. Using lessons learned from this project, jurisdictions will be better prepared to deal with their own unique urban challenges, positioning themselves to capture real-time availability and implement performance-based pricing.

Since the launch of a performance parking pilot by SF*park* in August 2011, theories set forth by the likes of Nobel Laureate William Vickrey and Donald Shoup related to adjusting the cost of on-street parking to establish an equilibrium between pricing and the availability of on-street parking have been at the forefront of managing on-street parking in major U.S. cities. Some cities, like Indianapolis, Cincinnati, and Chicago, used general occupancy trends to set rates in larger "zones." Others, however, followed San Francisco's lead, including Los Angeles, New York, Seattle, Boston, and Washington, D.C., exploring the marriage of parking rates to utilization in a more precise manner.

The approach used by the San Francisco Municipal Transportation Authority (SFMTA) to measure historical real-time parking occupancy relied upon the use of in-ground sensors, generally using at least one sensor per space (*1*). The SF*park* initiative, involving 8,200 on-street parking spaces, used 11,917 sensors, 346 repeaters, and 58 gateways, a ratio of 1.45 sensors per space (*2*, *3*). Parking occupancy studies in Los Angeles and Indianapolis had one-to-one sensor-to-space ratios in the study areas. New York City tested using at least two sensors per space (placing sensors every 5 to 8 ft), but found that cameras best captured occupancy data when spaces are not demarcated, as discussed below (*4*). In Washington, D.C., the District Department of Transportation (DDOT) estimated the capital cost to implement every one of the 18,000 metered spaces to be between \$2.25 million and \$4.8 million, depending on the specific technology. Further, additional annual operating costs could vary between \$1 million and \$3.2 million (*1*). DDOT determined this approach would be unsustainable on a large scale.

Cities have attempted other approaches to measure occupancy as well. Since 2011, the New York City Department of Transportation (NYCDOT) has used time-lapse photography-based technology to monitor parking, occupancy, and turnover. NYCDOT found time-lapse technology to be flexible and cost efficient, and it was readily available through off-the-shelf technology. The use of time-lapse photography required a manual review of footage and the help of NYCDOT technicians to install the cameras (*5*). The City of Seattle conducts an annual parking occupancy data collection effort covering

[1]District Department of Transportation, Washington, D.C.
[2]Kittelson & Associates, Inc., Washington, D.C.
[3]Parking and Mobility Solutions, Conduent, Chicago, IL
[4]Parking and Mobility Solutions, Conduent, Austin, TX

Corresponding Author:
Address correspondence to Soumya S. Dey: soumya.dey@dc.gov

all city neighborhoods with paid on-street parking. The data collection effort is intensive, and includes data from 23 neighborhoods during four weekday time bands (8.00 a.m. to 12.00 p.m., 12.00 p.m. to 3.00 p.m., 3.00 p.m. to 6.00 p.m., 6.00 p.m. to 8.00 p.m.), and a Saturday peak time period. The occupancy information is collected manually, and because of the significant resources required to collect the data manually, the City has worked to identify more efficient ways to collect on-street parking data (*6*). However, neither the time-lapse approach used by NYCDOT nor the manual approach used by the City of Seattle provide municipalities with real-time parking information or accurate duration data.

SF*park* more recently experimented with a no-sensor approach to provide real-time information. Between August 2011 and the end of 2013, SF*park* collected occupancy data through in-ground sensors. After implementing 13 demand-responsive rate adjustments using data from the in-ground sensors, SFMTA developed the Sensor Independent Rate Adjustment (SIRA) methodology to continue rate adjustments without the use of sensor technology. In developing the SIRA model, SFMTA chose to use a linear model which incorporated payment rate along with neighborhood type and day of the week as variables. SFMTA did not use a log-log model with the same variables because of the "inherent uncertainty in estimating parking occupancy rates using a model rather than parking sensors...when it is wrong, [the linear model] is more likely to set rates that are too low rather than too high" (*7*).

In addition to the limitations of sensor technology and available data to provide historical and real-time parking occupancy in a cost-effective manner, unique challenges are inherent in complex urban environments with multiple transport modes and a greater density of buildings and people. Prior to the installation of occupancy detection equipment in Washington, D.C., the project team understood that many urban challenges had to be overcome in San Francisco, Indianapolis, and New York City. In San Francisco, the in-ground sensors had to overcome difficulties with the presence of streetcars (*8*). Indianapolis experienced system-wide performance issues related to in-ground sensor battery life. In New York City, sensors installed in the Bronx had difficulty accurately capturing data on streets without demarcated parking spaces. Attempts to develop an algorithm to predict parking occupancy on non-demarcated blocks failed to provide predictive occupancy data with enough reliability to be used for public consumption. Lastly, anecdotal evidence suggests other cities have experienced battery issues due to inclement weather, especially shortened battery life in colder conditions.

Using lessons learned from San Francisco, New York City, Seattle, and Indianapolis, DDOT sought to develop a system that provides customers with real-time parking occupancy information while leveraging the collection of historical occupancy data to be used to adjust pricing in a more cost-effective manner. To do so, DDOT developed an innovative, "asset-lite" approach informed by other data sources, either to supplement or supplant sensor and meter payment data, and to paint a picture of occupancy that allows for accurate rate recommendations that help get motorists to their destination. To be successful, this asset-lite approach had to be compatible with the ever-changing and complex urban environment in which it is located.

This paper introduces the parkDC project, explores the urban challenges confronted by DDOT, discusses how the use of advanced analytics helped the project team overcome challenges related to technology, cost, and the urban environment, and provides lessons learned to better prepare other jurisdictions for addressing their own unique urban challenges.

PARKDC: Penn Quarter/Chinatown

DDOT is presently conducting a pilot study to implement demand-based pricing for on-street parking in the Penn Quarter and Chinatown neighborhoods in downtown Washington, D.C. As part of the study, DDOT is applying an "asset-lite" approach to occupancy detection and prediction that seeks to fuse data from multiple sources, including in-ground parking sensors, fixed and mobile cameras, computer vision, enforcement and citation issuance, and meter payment transactions. Occupancy and hourly meter pricing and policies are available to customers through mobile applications developed by DDOT and by private app developers using an open-source portal.

Through the program, DDOT seeks to improve the parking experience for customers, encourage the effective use of curbside space, and reduce downtown congestion associated with parking activity for personal vehicles, motorcoaches, and freight vehicles. By reviewing data before and after the evaluation, the study will measure the impact on parking occupancy, turnover, and transaction data. It includes intercept surveys; and collects and reviews field data such as parking search time and percentage of vehicles "circling" to find an open parking space. The project will provide DDOT with an evaluative tool to determine the most cost-effective technologies (the "minimum-viable solution"). Our focus in this paper is on how the asset-lite components of the sensor deployment and occupancy estimation strategy can achieve that minimum-viable solution.

The study area, represented in Figure 1, comprises three partially overlapping D.C. neighborhoods: Downtown, Chinatown, and Penn Quarter. The area is located in the heart of D.C., with the National Mall to

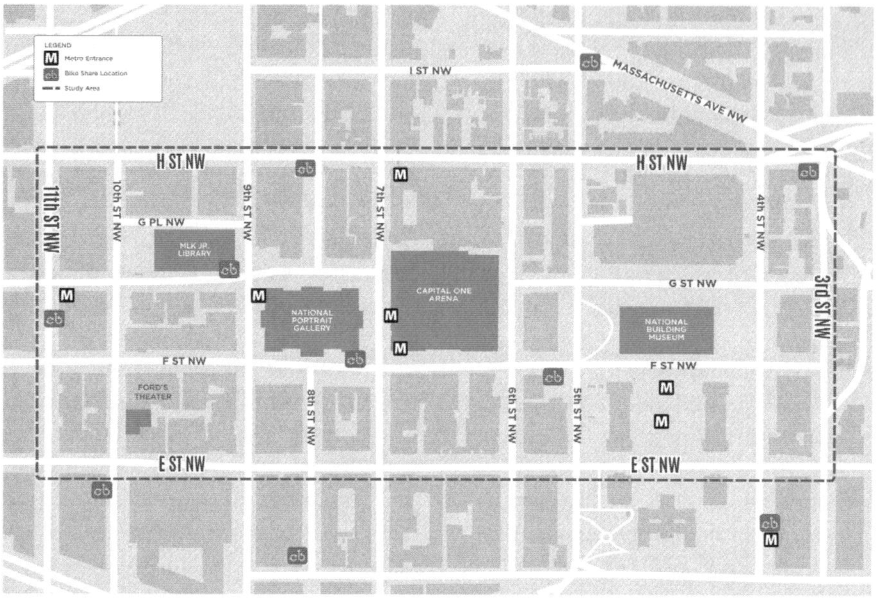

Figure 1. The Penn Quarter/Chinatown study area.

the south, the White House to the west, the Convention Center to the north, and Union Station to the east. Popular destinations in the area include the Capital One Arena, the National Portrait Gallery, Chinatown, and a vibrant commercial district, all creating a consistently high demand for parking.

From a transportation perspective, the study area includes 114 block faces (91 with parking meters), approximately 900 metered spaces, 30 loading zones, six Capital Bikeshare stations, two on-street car-sharing spaces, three Metrorail stations serving all regional Metrorail lines, Metrobus stops, and freeway-arterial interactions to the east of the study area.

Informed by NYCDOT's experience and a desire to provide customer convenience, DDOT transitioned from undemarcated multi-space meters to a demarcated pay-by-space environment (*9*). Under the new configuration, space markers identify parking spaces on each block. Fixing the number of spaces and where cars can park along the block improves the accuracy of occupancy detection technologies, in part by ensuring a fixed supply throughout the day, and increases the granularity of transaction and enforcement data. Studies have shown that pay-by-space environments may be more efficient than unmarked spaces when occupancy is high (*9*).

Urban Challenges

Sensor Communication

To report real-time parking occupancy information accurately, in-ground sensors must report their status (occupied or not occupied) to a nearby router. This requires sensors to communicate with the router, and the router to communicate with a central server via a mobile phone network or a wired connection. For the pilot project, the mobile phone network was used to minimize the capital costs related to the project. However, several challenges unique to urban environments required additional set up and maintenance to ensure accurate and timely communications from the sensors to the back office:

- Jamming devices: During testing of the installed in-ground sensors, a sensor vendor identified issues with jamming devices being used in adjacent government buildings.
- Intense mobile signal: What was believed to be a very strong mobile signal being emitted from the Capital One Arena located within the parkDC: Penn Quarter/Chinatown study area.
- Metrobus and Metrorail trains: Additional issues were identified related to Metrobus vehicles indicating false positives, and Metrorail trains below the roadway also impacting sensor readings, similar to issues San Francisco faced with streetcars.
- Installation: Further communications issues arose when a modem was found to be failing due to water damage caused by the over-tightening of a screw in the communication housing during installation. In this case, the modem failure took approximately 30 sensors networked through the modem offline while the modem was replaced.

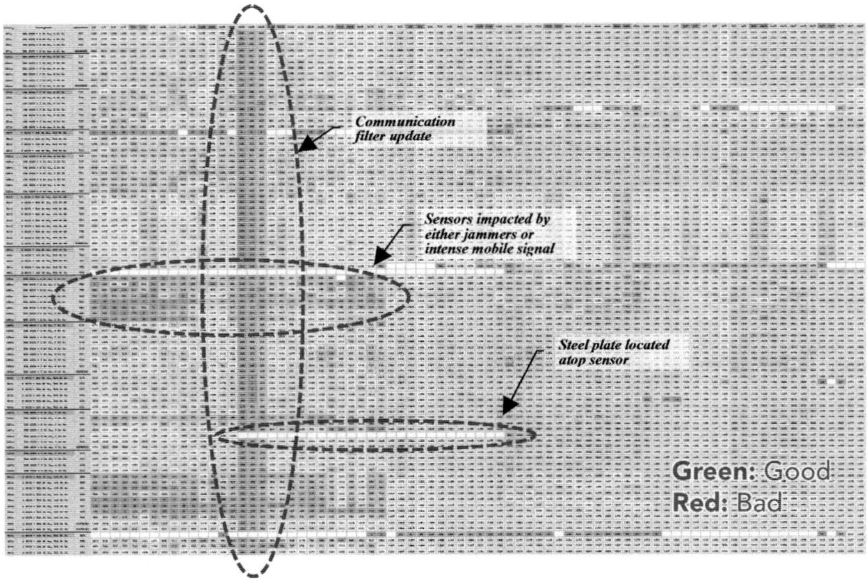

Figure 2. Testing of installed in-ground sensors.

- Issues in surrounding uses: A small fire from a nearby construction site rendered a cellular modem enclosure obsolete, eliminating the ability to receive data from three blocks for an extended period.

When sensors fail, the malfunction may not be communicated immediately. By proactively studying data patterns, the project team could predict and communicate likely sensor failures to vendors, allowing for the problems to be addressed more quickly.

One sensor vendor detected some issues by reconfiguring the sensors' "health" communication from once every 24 hours to once every minute, thereby enabling the vendor to deduce that sensor messages were not being received due to signal interference. The vendor overcame the jammers and intense mobile signal by updating the filters on the communications gateways, effectively improving the communication quality and reducing interference from federal agencies and other facilities. Back-office system modifications (revised algorithms) were required to address the impacts from buses and trains on the magnetometers.

Figure 2 shows the output of a daily "heartbeat" test performed by the sensor vendor. The rows represent individual sensors, and the columns indicate multiple daily tests. As the number of "heartbeats" or communication pings from the sensors to the communication equipment gets close the desired range, the conditional formatting adjusts from red to green (red representing too few "heartbeats" received and green representing an appropriate number of "heartbeats"). Where there were no pings or there were obstructions to parking spaces (like street closures or construction), only blank white cells are shown. An early communications filter update is easily identifiable in the annotated figure; most of the sensors had fewer heartbeats received, as indicated in "red," at this time. Looking to the days following the communications update (right of the update), it is clear the filter application improved communications, reducing the number of "red" sensor indicators to the right of the update and increasing the sensors with optimal, or "green," indicators moving forward.

Whenever data accuracy is in question due to street closures, communication issues, and/or other complications, the data from the impacted sensors are excluded from the analysis in determining occupancy and deriving recommendations for price changes. Using a combination of data from other sources, including mobile payments, historical occupancy information, and citations, and validating against similar data from adjacent blocks, the data gaps can be filled.

Limited Space, Many Demands

Portable CCTV Cameras. At the outset of the project, six trailers with cameras were moved on a weekly basis throughout the pilot area using the Whittle Index (*10*) restless multi-armed bandit approach to guide the placement of cameras to capture information on 58 blocks for use as baseline data. Because single cameras did not cover many spaces, multiple cameras were often required to cover a full block face. Up to four cameras were mounted on each trailer, and while the trailers were intended to have minimal impact in the urban environment, with dimensions around the size of a compact car (12.5 ft length × 7 ft width, with a height of 8 ft with the camera retracted or 30 ft when the camera was fully extended), the trailers were often still disruptive.

Moving the trailers each week was manually intensive and could take several hours for each trailer. At each location, a decision was made to place the trailer either on the sidewalk—potentially obstructing sidewalk activity—or in a parking space, reducing revenue opportunities, and risking running counter to the goals of the overall project. Further, in many locations, peak-hour restrictions and unexpected construction permits restricted the ability to use camera trailers. Other concerns were raised, including a local organization with a major meeting scheduled that did not want an unsightly camera placed near the entrance to their downtown office building with VIPs arriving and federal government security concerns.

A methodology was developed for guiding the installation of mobile CCTV trailers to optimize accuracy. It was imperative for operators to set up the cameras properly to ensure necessary data capture and analysis using computer vision. Obstructions like large vehicles and trees can impair the accuracy of the counts and were factored into the analysis. During installation, attempts were made to minimize these obstructions. Further, the computer vision algorithms were modified to address the effect of shadows and the potential of camera shake due to wind or passing large vehicles. This required the computer vision algorithms to be individually fine-tuned for each installation.

Demands for Pole and Traffic Cabinet Space. Additional hardware to support the in-ground sensors necessitated placement of combined signal and light poles at intersections, as did the installation of Bluetooth sensors at 59 intersections in the study area to measure the number of vehicles cruising for parking. With space on these poles needed for signal and illumination hardware, along with other hardware, the space available on poles is at a premium. Field visits assessed pole space availability when selecting where to hang equipment at individual intersections.

Similarly, because of insufficient space in the traffic signal cabinets to house a variety of computing, networking, power management, data storage, and communication electronics for the in-ground sensors, a separate enclosure was required to house the cellular modems and access point controllers. The enclosures also required a power supply, and when solar power was infeasible, a connection to street-light poles was required. This required additional coordination, including the presence of DDOT signal technicians to oversee the electrical connections. This further necessitated the coordination of bucket trucks to install the gateways, repeaters, and other communications infrastructure. This will continue to be a challenge as various "smart city" approaches encourage the expansion of sensing and camera technologies.

Demarcation. As noted above, to allow for the collection of revenue and occupancy data at a space level and to improve the customer experience, the decision was made to convert the study area from a pay-and-display configuration (where customers have to return to their vehicle to display a receipt after paying) to a pay-by-space configuration (customers enter the space number at the meter and do not have to return to their vehicle) (*9*). While studies show that demarcation may improve available curbside supply during peak periods, going from a non-demarcated to a demarcated environment requires work to implement. This includes having to install and maintain new space marker signage (or paint, if a city chooses that approach), updates to meter and enforcement software, public outreach to explain the change, and field quality control reviews to ensure the markers and software denote the correct spaces.

On-Street Activity Affecting the Data Collection

A variety of on-street activities, including paving and restriping, construction, emergency utility work, and road closures for other activities, may affect data collection:

- Paving and restriping can change the roadway configuration, including removing or relocating on-street parking spaces. While the project team attempted to place a moratorium on changes within the study area, the rapidly changing urban landscape, including building developments that were not going to be held up, still necessitated changes.
- Construction, whether on the road itself or an adjacent building, can require the closure of on-street parking, which prevents the collection of parking data. The data reviewed for the fourth price change identified 26 blockfaces (29% of the study area) with construction activity, shown in Figure 3, and therefore no real-time occupancy data. In addition to smaller construction projects requiring the project team to identify the location, duration, and extent of the impact to address the data discrepancy from the affected in-ground sensors, other larger construction projects also affected parking. However, the effects of these larger projects, such as the closure of the 400

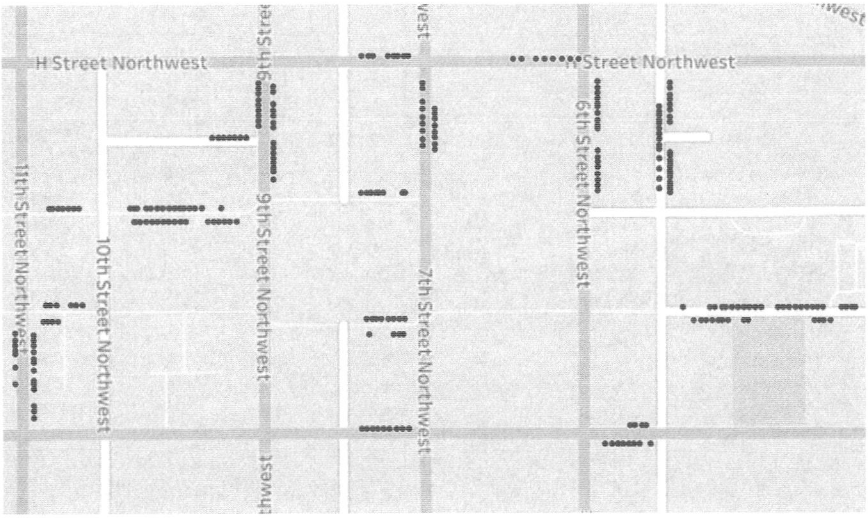

Figure 3. Blockfaces with construction permits and their associated on-street parking spaces during the fourth price change.

block of E Street, NW due to the construction of the National Law Enforcement Museum, were easier to deal with.

- Steel construction plates: Due to in-street utility construction activity, steel construction plates were found to be covering the sensors, creating false positive readings. Additional coordination was then necessary to get the steel plates moved more expeditiously, or the data from those sensors was temporarily removed from the readings.
- Curbside management: Routine curbside activities, including reserving parking spaces for an extended period for loading and unloading, construction, street festivals, or other activities, can regularly put on-street parking spaces out of service for several hours, days, or weeks in some cases.

It is difficult to prepare for these disruptions when the project team has no indication of when they will occur. DDOT has an online Transportation Online Permitting System (TOPS) which enables D.C. residents and businesses to apply for permits and provides the ability for DDOT to monitor the requests internally. At the onset of the project, alerts via email were set up to receive an automatic message for any request in the study area. However, the abundance of requests quickly became overwhelming, and the system was modified to pass along only requests involving the closure of a parking space in the study area.

Special Events

The Capital One Arena is a sports and entertainment arena located in the center of the study area, with a capacity around 20,000 people depending on the event and the configuration of the arena to host the event. Events at the arena can cause additional roadway congestion within the parkDC study area. A focused examination of events held at the Capital One Arena was undertaken by the study team to determine the impact the events might have on parking use in the study area, and to determine whether event-based pricing would have performance benefits.

As shown in Figure 4, the number of events at the Capital One Arena varies seasonally. From October 2016 through March 2017, events at the Capital One Arena occurred almost every day of the week. By comparison, in June, July, and August 2017, the Capital One Arena had six, five, and seven days, respectively, with scheduled events. While this provides a consistent and robust set of event data from November 2016 to March 2017, it does not allow for the collection of a robust sample size to calculate a rate recommendation for non-event days. Conversely, the opposite is true from June 2017 through August 2017, with plenty of days to collect non-event day data, but not enough days to collect special-event data. Historical event calendars for the Capital One Arena revealed this seasonal trend of having a high frequency of events from November through March to be consistent, as was having a low frequency of events from June through August. Further, the project team wanted to tread lightly in its approach to event pricing since any attempts to adjust prices for events could be perceived by the general public as price gouging, especially since the likelihood of shifting demand was low.

Despite the concerns about sample size, parking occupancy information for event and non-event days was reviewed, as demonstrated by Figure 5. The charts show

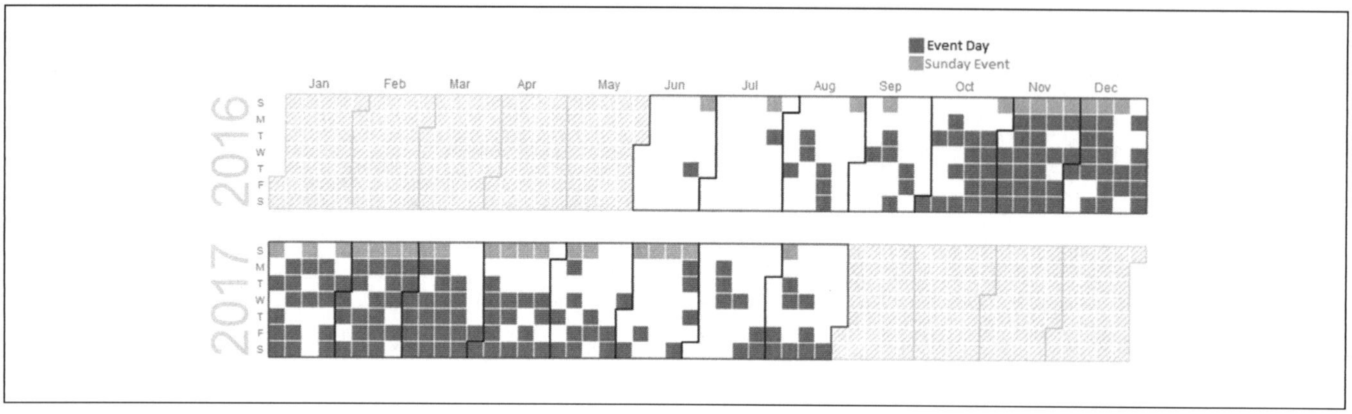

Figure 4. Capital One Arena events by month.

occupancy by day on the x-axis, and by day of the week and time of day on the y-axis. In July and August (towards the top of each chart), occupancy for event and non-event days is similar. However, when the Capital One Arena begins holding more events in the fall, occupancy for event days is noticeably higher than on non-event days. While this is most significant during the evening time periods, occupancy during non-evening time periods also tends to be higher on event days than non-event days. The small sample size, however, makes it difficult to confirm the evening impact at a statistically significant level.

Figure 6 reflects changes in occupancy on different blocks in the study area on event and non-event days during the evening time segment (4.00 p.m. to 10.00 p.m.). As shown, the east area of the study area is the most affected by events. Parking in this area predominantly serves large government institutions (the Government Accountability Office headquarters, and judicial buildings) that have higher demand during the daytime. Consequently, more parking tends to be available in the eastern portion of the study area in the evening when event demand is highest.

Based on the reviewed event data, the project team determined that the seasonal pricing changes largely addressed event attendance. The approach does not increase the error percentage in the rate-adjustment calculations because the number of events varies consistently by season. In addition, events at the Capital One Arena are of different sizes and magnitudes, making pricing adjustments difficult since changes would need to be tied into the expected magnitude of each event and monitored on a near daily basis. Further, specialized pricing for event days in the study area would have been difficult to communicate to the public.

Broader Mobility Affecting Demand

As the urban core of the Washington, D.C., metropolitan area, situations affecting the transportation system in the study area can reverberate region-wide, and have especially large impacts in the District. Between June 4, 2016 and June 25, 2017, Washington Metropolitan Area Transit Authority (WMATA) used 16 "Safety Surges" as part of its SafeTrack program to complete maintenance work on the Metrorail system. These "Safety Surges" reduced capacity on the Metrorail system by closing different segments of the system during each surge. To address the expected added roadway congestion during these surges, DDOT expanded the hours of operation for peak-hour restricted parking, which necessitated coordination to adjust signage and data collection to accommodate the changes.

The project team reviewed Metrorail data to identify if SafeTrack had an impact on ridership within the study area, and if any correlation between SafeTrack and changes in parking occupancy could be identified. Figure 7 displays the daily entries and exits at Metro stations within the study area since the time of the initial price change. As can be seen, the number of entries and exits remained relatively stable after the first price change, apart from the large increase in January 2017, which corresponds with the Women's March on Washington the day after the Presidential Inauguration (many of the Metro stations in the study area were closed for portions of the day during the inauguration, lowering their ridership totals). When looking at this data aggregated by month on weekdays with outliers (Presidential Inauguration, holidays, etc.) removed, as shown in Figure 8, it becomes evident that Metro use within the study area has increased since the implementation of the first price change. This compares with a ridership decrease of about 12% on the Metrorail system as a whole in a similar time period (*11*), which is largely attributed to SafeTrack. While the increase in Metrorail use within the study area cannot be attributed to the implementation of price changes for on-street parking, the Metrorail ridership numbers are moving in the right direction for that to have occurred.

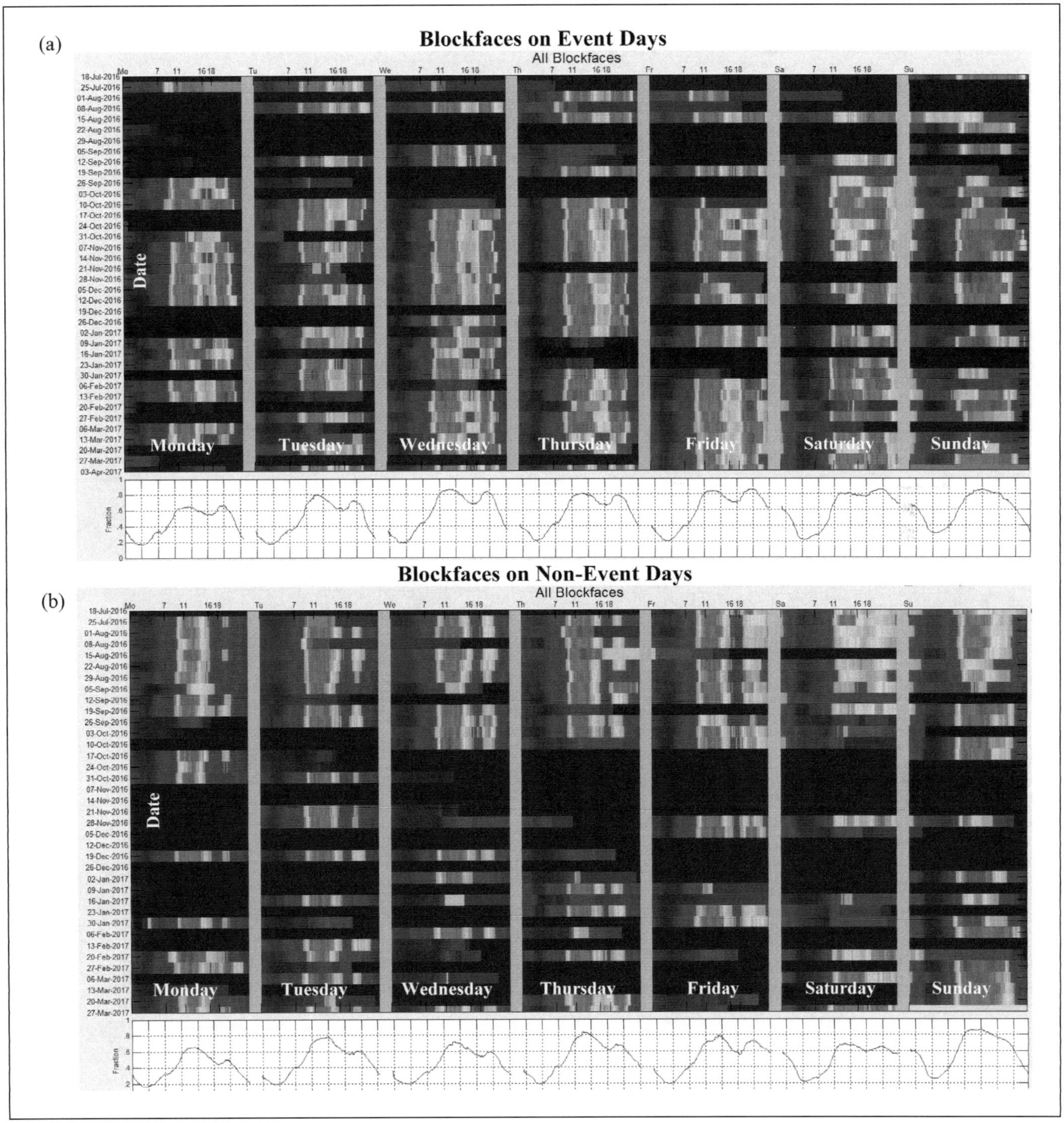

Figure 5. Parking occupancy during event days (top) and non-event days (bottom): (*a*) blockfaces on event days; (*b*) blockfaces on non-event days.

Parking Users

The types of users of on-street parking spaces vary, with commuters, tourists, and retail shoppers comprising most of the curbside parkers. Within these groups, special circumstances, occupations, or employers allow for the use of placards, such as disabled placards, government-vehicle placards, or delivery vehicles. Many of these placards allow for vehicles to park in certain—or sometimes all—on-street parking spaces without paying and for longer periods of time than other vehicles. Because these

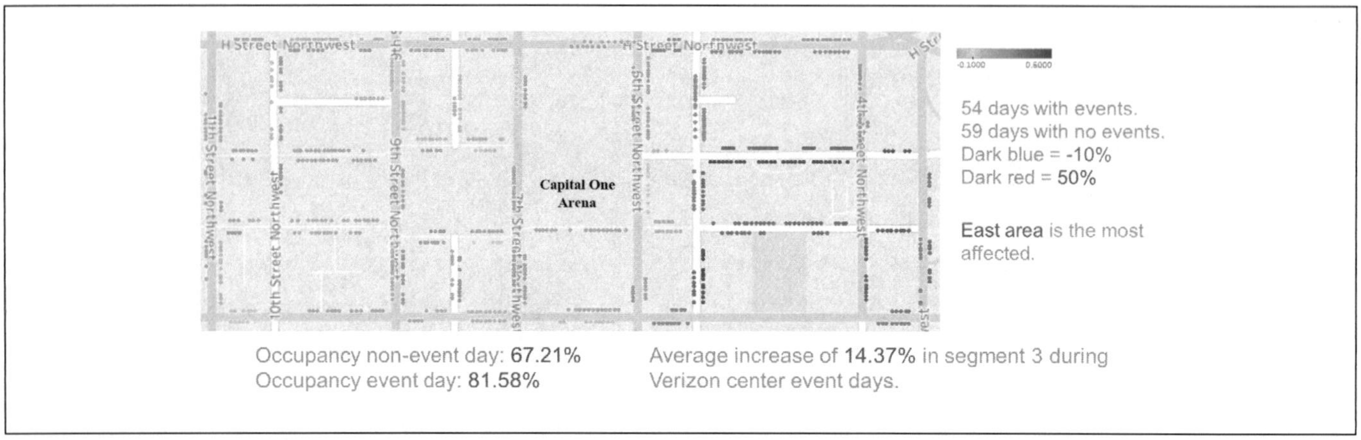

Figure 6. Difference in occupancy: event days versus non-event days on weekday evenings (6.00 p.m.–10.00 p.m.).

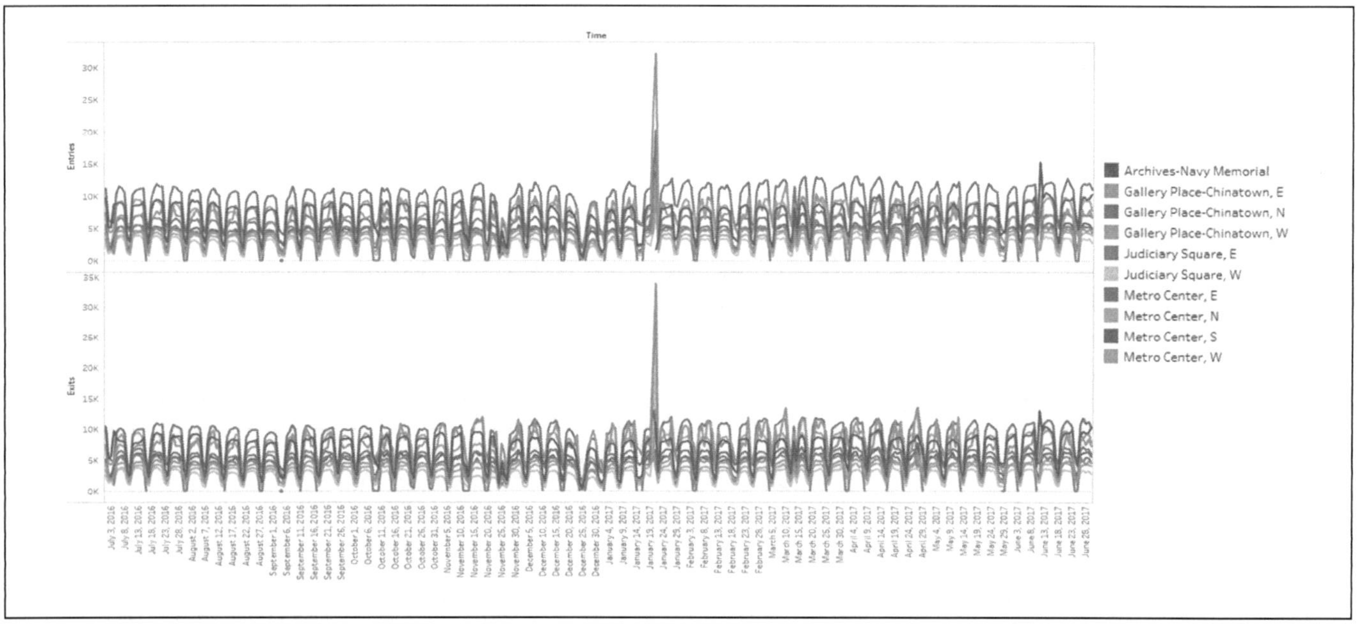

Figure 7. Entries and exits per day at study area Metro stations.

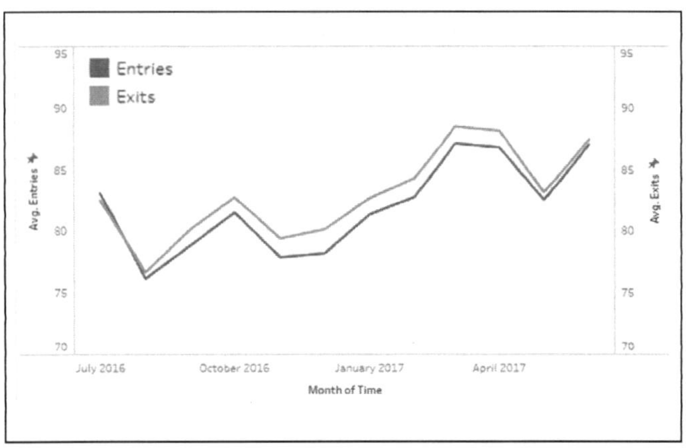

Figure 8. Average weekday Metro entries and exits by month (no outliers).

vehicles are insensitive to price changes, other strategies need to be employed to monitor and shift these parking users. This has included outreach to law enforcement and other agencies, as noted in the section "Interagency Coordination" (below), exploring alternative pricing strategies for loading zones to discourage misuse, and implementing the "red-top" meter program to reserve parking for disabled placard holders but to require payment and impose time limits for that use (*12*).

Installation

Lastly, there were several challenges associated with installing sensors, some elements being particular to cities and others (like weather) having more general application:

- Closure of on-street parking: The restriction of parking on D.C. streets to install items like in-ground sensors requires significant permitting and public notification processes compared with many other cities. Further coordination was necessary to avoid conflict with events at the Capital One Arena. Despite closures being announced in advance, along with posted signs and cones placed on the street, drivers continued to park in the closed on-street parking spaces. This further delayed the installation of the in-ground sensors, and required securing enforcement officers and tow trucks. However, in some cases the vehicles belonged to government agencies that could not be towed. In these cases, personnel were required to install sensors in non-peak hours.
- Installation conditions and noise restrictions: Less than ideal weather conditions also required work during off hours, including during the nighttime. Nighttime noise restrictions further reduced the ability of the installation team to install the sensors with the exception of small windows of time between busy daytime hours and the overnight noise restriction hours. Due to a particularly wet spring in the D.C. area, the conditions during installation were often wet, requiring the use of heaters to dry holes and ensure the epoxy formed the necessary bond. Occasionally work could not be completed during the week scheduled because of competing construction work or other projects, necessitating additional trips and coordination activities.

Interagency Coordination

Because varying agencies develop parking regulations and policy (DDOT), enforce parking regulations (Department of Public Works, or DPW), adjudication (Department of Motor Vehicles, or DMV) and supplement enforcement (the Metropolitan Police Department, or MPD), multiple municipal agencies needed to be involved to move forward with tasks like revising loading zone pricing and changing the payment structure from "pay-and-display" to "pay-by-space." In changing the payment structure, handheld devices used by the enforcement officers had to be updated, and officers required training on the change to demarcation; efforts funded by DDOT, but carried out by DPW.

Due to construction and other on-street activities, WMATA occasionally moves bus stops. When this happens, coordination helps ensure demarcated parking spaces are taken offline, and sensors are removed from reporting real-time occupancy information.

The presence of federal government buildings and related security measures in Washington, D.C. adds additional agencies for DDOT to coordinate with. The installed assets, including cameras and gateways, were marked with stickers to inform non-District personnel, especially security agencies located in Washington, D.C., of their intended use. DDOT's logo and phone number were also placed on the stickers to allow for verification.

Prior to the implementation of price changes, observations indicated many law enforcement vehicles parking within the eastern third of the study area. Users of these vehicles are exempted from paying for parking, which, in addition to being a missed revenue opportunity, does not allow for occupancy data to be collected based on payment data. As part of this project, DDOT requested MPD and the Federal Bureau of Investigation (FBI) to encourage officers from both agencies to limit on-street parking of their agencies' vehicles to blocks already designated for government vehicles.

Conclusion

Sustainable, asset-lite approaches have the potential to become the state-of-the-practice for developing real-time availability of on-street parking in an urban environment. By consulting with data scientists and using advanced analytics, along with sound project management principles, changes to the rapidly changing urban environment can be accommodated and disruptions to the project can be minimized. Using lessons learned from this project, jurisdictions will be better prepared to deal with their own unique urban challenges, and position themselves to implement real-time availability information and performance-based pricing.

Lessons Learned

Allow for Flexibility. The ever-changing urban environment makes it almost impossible to expect the baseline conditions to remain constant throughout a project. Despite DDOT's best efforts to identify a controlled area and put in place moratoriums on construction and curbside use changes, factors outside of DDOT's control necessitated flexibility and adjustments to the study area. In some cases, these changes were viewed as opportunities to collect additional data and conduct additional analysis on the impacts these changes had on parking occupancy and mobility within the study area.

Start Monitoring Early. Early issues with sensor installation and communication were identified and addressed due to the ongoing "heartbeat" monitoring of the sensors taking place. Furthermore, working with vendors early, before installation, helps address issues related to interference, jamming signals, or communication issues. This monitoring also extends to on-street activities that will affect the ability to collect data.

Consider Special Events. All major cities have unique special events that influence demand for on-street parking in different ways. Considering how these events operate, and how the parking ecosystem can be designed to accommodate them, can help minimize customer complaints and improve system efficiency.

Communicate Early and Often. Inter- and intra-agency communication have been imperative as the project has moved forward. Conversations within DDOT, and externally with other agencies and neighborhood groups (the business improvement district, neighborhood associations, and the local governing Advisory Neighborhood Commissions), have helped educate the public, inform agencies of impending changes, and flag issues to the project team.

Explore Policy Solutions. Some of the challenges with parking users cannot be addressed by pricing or enforcement alone. DDOT's policy changes around loading zone pricing (extant when the pilot began) and red-top meters have reserved access for certain curbside users to address larger system and user needs. Understanding how policy changes occur, especially if ordinances or council or legislative action is required, and providing advance lead time to change regulations, especially if governmental bodies need to be engaged, is also important.

Budget for Sensor Relocation. Despite DDOT's best efforts to avoid disruptions to the installed sensors, inevitably sensors were affected by construction or roadway restriping. Having budget previously allocated would have made the relocation of sensors easier.

Acknowledgments

This project was funded in part by the Federal Highway Administration's Value Pricing Grant. The authors acknowledge the contributions of Evian Patterson of DDOT, Mark Derrick of Conduent, and Brandon Nevers of Kittelson & Associates, Inc.

Author Contributions

The authors confirm contribution to the paper as follows: study conception and design: Soumya Dey, Stephanie Dock, Benito Pérez, Matt Darst, Eduardo Cardenas, Alek Pochowski; data collection: Matt Darst, Eduardo Cardenas, Alek Pochowski, Meredyth Sanders; analysis and interpretation of results: Soumya Dey, Stephanie Dock, Benito Pérez, Matt Darst, Eduardo Cardenas, Alek Pochowski, Meredyth Sanders; draft manuscript preparation: Stephanie Dock, Alek Pochowski, Meredyth Sanders. All authors reviewed the results and approved the final version of the manuscript.

References

1. Dey, S. "Asset Lite" Payment Options and Occupancy Detection for Metered Curbside Parking. *ITE Journal*, Vol. 86, No.4, 2014, pp. 29–35.
2. San Francisco Municipal Transportation Agency. *Parking Sensor Technology Performance Evaluation.* 2014. http://sfpark.org/wp-content/uploads/2014/06/docs_sensorevaluation.pdf. Accessed July 20, 2015.
3. San Francisco Municipal Transportation Agency. *Parking Sensor Data Guide.* 2013. http://sfpark.org/wp-content/uploads/2014/06/docs_sensordata.pdf. Accessed July 20, 2015.
4. Kazis, N. *City Tests out Parking Sensors, but So Far Just for Space-Finding App.* 2012. http://www.streetsblog.org/2012/01/18/city-tests-out-parking-sensors-but-so-far-just-for-space-finding-app/. Accessed July 20, 2015.
5. Blakely, M., and N. Gray. Time-Lapse Cameras Measure Street Parking Demand. *ITE Journal*, Vol. 83, No. 9, 2013, pp. 36–39.
6. Kaufman, M., M. Formanack, J. Gray, and R. Weinberger. *Contemporary Approaches to Parking Pricing: A Primer.* Federal Highway Administration, Washington, D.C., 2012.
7. San Francisco Municipal Transportation Agency. *Sensor Independent Rate Adjustments (SIRA) Methodology and Implementation Plan.* 2014. http://sfpark.org/wp-content/uploads/2014/05/SIRA-methodology-and-implementation-plan_2014_05-14.pdf. Accessed July 20, 2015.
8. JVH. *SF Park Is a Success, Failure, Neither, Both???* 2013. http://www.parkingtoday.com/blog/2013/12/sf-park-is-a-success-failure-neither-both/. Accessed July 28, 2016.
9. Dey, S. S., C. R. Dance, M. Darst, S. Dock, T. Silander, and A. Pochowski. To Demarcate or Not to Demarcate: Analysis of Marked versus Unmarked On-Street Parking Efficiency. *Transportation Research Record: Journal of the Transportation Research Board*, 2016. 2562: 18–27.
10. Whittle, P. Restless Bandits: Activity Allocation in a Changing World. *Journal of Applied Probability*, Vol. 25A, 1988, pp. 287–298.
11. Smith, Max. *Metro Ridership Drops 12 Percent; $125 Million Revenue Shortfall Projected.* 2017. http://wtop.com/tracking-metro-24-7/2017/02/metro-ridership-drops-12-percent-125-million-revenue-shortfall-projected. Accessed July 28, 2017.
12. District Department of Transportation. *Disability Parking.* https://ddot.dc.gov/page/disability-parking. Accessed July 31, 2017.

The Standing Committee on Transportation Issues in Major Cities (ABE30) peer-reviewed this paper (18-04926).

Article

TRR
JOURNAL OF THE TRANSPORTATION RESEARCH BOARD

Transportation Research Record
2018, Vol. 2672(7) 69–81
© National Academy of Sciences:
Transportation Research Board 2018
Article reuse guidelines:
sagepub.com/journals-permissions
DOI: 10.1177/0361198118791863
journals.sagepub.com/home/trr
SAGE

Developing a Toolkit to Improve Transportation Safety on Indian Reservations

Sahima Nazneen[1], Trenna L. Terrill[2], and Khaled Ksaibati[3]

Abstract

Indian reservations hold the highest rate of crashes that lead to fatal and incapacitating injuries across the United States. Limited resources, absence of coordination across jurisdictions, and limited crash data make it difficult for tribes to reduce the number of these fatal and serious crashes. A safety toolkit was developed in this work to identify the high-risk crash locations and determine low-cost safety improvement countermeasures. The toolkit serves as a guideline and provides information, field examples, and resources in key topic areas to lead the effort to improve safety via the use of a five-step methodology. These five steps are, namely, compile data and crash data analysis, Level I field evaluation, combined ranking, Level II field evaluation to identify countermeasures, and benefit–cost analysis. The objective of the toolkit is to assist tribes to reduce the number of fatal and serious crashes, and provides flexibility for the tribes to utilize the methodology compatible with the onsite data, preferences of the tribes, and other factors to meet demands. The methodology described was successfully implemented on the Wind River Indian Reservation, Standing Rock Sioux Tribe, Sisseton Wahpeton Oyate Tribe, and the Yankton Sioux Tribe and showed great success in identifying high-crash locations.

Indian reservations experience the highest rate of fatalities and serious crashes among all ethnicities across the United States. According to NHTSA (2009), American Indians experienced the highest fatality rate per 100,000 populations in 2006 (*1*). Poor road conditions associated with rustic roadways, coupled with risky driving behaviors such as speeding and impaired driving significantly increase the chances of fatal and serious crashes on these roadways. Moreover, lack of resources, an absence of coordination across jurisdictions, and limited crash data contribute to make it difficult for tribes to implement an effective road safety program (*2*).

In order to address these challenges, a safety toolkit was developed by the Wyoming Technology Transfer Center, Local Technical Assistance Program (WYT^2/LTAP) with funding from FHWA. The goal of the safety toolkit is to assist tribes in determining high-risk crash locations and identify potential low-cost countermeasures. This tool kit provides Indian Nations with the opportunity to utilize the funds for safety improvements on their roadway systems efficiently. The methodology described in the toolkit was implemented initially on the Wind River Indian Reservation (WRIR). After the successful implementation of the initial project, the program was expanded to the Standing Rock Sioux Tribe (SRST), the Sisseton Wahpeton Oyate Tribe (SWOT), and the Yankton Sioux Tribe (YST). It showed great success in reducing the number of fatal and incapacitating injuries occurring on those reservations roadways. However, despite the success of the methodology, challenges and difficulties were faced while implementing the five-step road safety process on these roadways, which indicated that a single methodology might not work for all areas. Consequently, the toolkit was updated to provide flexibility for the tribes to utilize the methodology compatible with available onsite data, the preference of the tribes, and other factors, to meet demands.

Extensive coordination and collaboration among the tribes and government agencies is essential for the successful implementation of the toolkit. WYT^2/LTAP and other technical assistance program centers were able to

[1]Civil and Architectural Engineering Department, University of Wyoming, Laramie, WY
[2]Wyoming Technology Transfer Center/LTAP, University of Wyoming, Laramie, WY
[3]Wyoming Technology Transfer Center, University of Wyoming, Laramie, WY

Corresponding Author:
Address correspondence to Khaled Ksaibati: khaled@uwyo.edu

provide the necessary support to facilitate implementation on the Indian reservations.

Objective

The objective of this paper is to describe the safety toolkit developed to determine high-risk locations and identify low-cost improvements on Indian reservation roadways across the United States. This toolkit considers the unique challenges tribal communities face. The five-step methodology used by the toolkit serves as a template for the tribes to develop their own safety plan.

Background

Three hundred twenty-six Indian reservations, home to 566 federally recognized Indian tribes, have 145,000 mi of roadways under the jurisdiction of the Bureau of Indian Affairs (BIA), the tribes, and state and local governments (*3*).These areas have witnessed a higher number of crashes over the years than other areas across the United States. Adult vehicle-related fatality rates for American Indians are 1.5 times higher than that of other ethnicities (*4*). A National Tribal Transportation Safety Summit report stated that impaired driving and the lack of use of seat belts/child safety seats are the highest concerns among the many safety issues on reservation roadways (*5*). Other issues that contribute to the high fatal crash rates on Indian reservations could be attributed to the condition of their roadways. According to condition and performance ratings reported by FHWA, about 83% of the total BIA and tribally owned roads were not in an acceptable condition in 2015 (*6*).

FHWA, NHTSA, BIA, the Indian Health Service, tribes, and states have made collaborative efforts to address the causes and find solutions to this disparity in crash, injury, and death rates in the Indian reservations (*7*). However, issues such as variation in data collection and analysis practices, data processing capabilities, and the unique nature of American Indian/Alaskan Natives (AI/AN) areas, and so forth, hinder efforts to ascertain priority areas. According to a study on reservations in South Dakota, approximately 64% of crashes were underreported in 2005 due to inadequate tribal law enforcement capacity, standardization issues of crash reports, and the relationship between the state and tribes (*8*). These issues also acted as barriers to addressing roadway the conditions contributing to crashes on the reservations.

WYT^2/LTAP developed the Wyoming Rural Road Safety Program (WRRSP), funded by the Mountain-Plains Consortium and Wyoming Department of Transportation (WYDOT) in cooperation with FHWA, to identify high-risk rural locations and develop a strategy to obtain funding to reduce crashes on the riskiest segments (*9*). As a derivative of the WRRSP, the WYT^2/LTAP developed a safety toolkit for tribal communities to improve safety on their roadways. As Indian reservation roads are similar to rural local roads, modifying the WRRSP to fit the needs of Indian reservation serves Indian Nations with the opportunity to identify low-cost safety improvements and then apply for and allocate funding for these improvements.

What is in the Toolkit?

This toolkit is geared to be used as a guideline and a source of information, field examples, and resources on key topic areas to lead the effort to improve safety through the use of a five-step methodology. This methodology mainly focuses on roadway characteristics rather than the behavioral aspects of crashes. Professional examples have also been included throughout to provide instances of successes and struggles in implementing the toolkit. The five steps of the methodology are:

1. Compile data and crash data analysis.
2. Level I field evaluation.
3. Combined ranking.
4. Level II field evaluation to identify countermeasures.
5. Benefit–cost analysis.

The methodology that was developed and implemented in the toolkit incorporated data sources and analysis tools available for tribes. The five steps incorporated in the tool kit were directly tailored to the specific challenges facing tribal transportation departments. These issues include:

- How to deal with incomplete datasets: Most tribes do not have complete data sets to conduct safety evaluations. The tool kit describes how all sources should be considered, and to obtain and combine crash data from many sources including state agencies, BIA officers, and local transportation agencies. As tribes are not obliged to report crashes, it is possible that crashes have been omitted from the crash databases maintained by state agencies.
- How to deal with pedestrian crashes: Pedestrian-related crashes are higher on tribal roads than other roadways. This is due to tribal members tending to walk more than the rest of the population. The tool kit tackles this issue in the first step, dealing with crash data analysis, as well as in the second and the fourth steps, which concentrated on identifying the conditions of pedestrian facilities and the benefits of implementing proper signage and pavement markings.

- How to implement safety countermeasures: The fourth step in the tool kit provides tribal members with the ability to select safety countermeasures consistent with what they consider fits with their communities.
- How to estimate the cost effectiveness of improvements: The benefit–cost analysis in the tool kit incorporated the reality that tribes tend to use their own workforce when implementing safety improvement measures.

The implementation of these steps is succinctly described in the following subsections.

How to Use the Toolkit

The first step of the toolkit compiles recorded as well as anecdotal crash data and analyzes them. This step is necessary to understand the factors contributing to high-risk crash locations. Once the crash ranking has been completed, the second step called the Level I field evaluation begins. This is an effective way to identify the most accident prone sites, and these sites are chosen for safety improvement. In Step 3 (combined ranking), crash locations are narrowed down to a list of high-priority locations for implementing countermeasures. Later, a Level II field evaluation is performed on each of the routes selected to determine the potential countermeasures. In order to maximize resources, the Level I field evaluation may be performed at the same time as the Level II field evaluation. Economic analysis is required when more than one countermeasure has been identified for a site, or when funds are insufficient to implement actions in all high-risk locations. After the countermeasures have been selected, the benefits and costs of the implementation can be evaluated. The following sections present detailed information about the toolkit, its tools, examples, guidance, and resources associated with it.

Compile Data and Crash Data Analysis

The first step in the safety evaluation toolkit is to compile the crash data and analyze them to determine high-risk crash locations. The objective of this section is to guide users in performing a complete crash analysis as well as to provide guidance to find the resources to aid in the data accumulation process. Issues such as the types of data and their sources need to be known beforehand to procure accurate and complete crash data.

Data Examples

Anecdotal Data. Anecdotal data are collected in an informal way, which includes phone calls from tribal members, community member survey results, local staff and police knowledge, and so forth. Anecdotal data cover a range of perspectives about potential safety issues as well as prospective countermeasures.

Quantitative Data. Quantitative data include information from police reports, crash data, traffic volume data, and roadway characteristics data. Crash data can be collected from local and state crash databases, as well as local police crash reports, NHTSA and its Fatality Analysis Reporting System. However, as sovereign nations, tribes are not obliged to submit their crash reports to the state agency; they may need assurance that the collected data will only be used for improving traffic safety. Crash rates can also be used as an effective tool to compare safety performance with comparable traffic volume and roadway characteristics elsewhere. Traffic volume data such as average annual daily traffic (AADT), vehicles miles traveled, major and minor street AADT can be acquired from state departments of transportation (DOTs). The handbook *Simplified Practice for Traffic Studies* provides information about collecting traffic volume data if none are available.

Data from Existing Plans, Documents, or Other Agencies

In addition to data, documents, information, and assistance from a variety of organizations and agencies can be referenced and enlisted as support for a crash data analysis. For example, the Strategic Highway Safety Plan (SHSP) provides a systematic approach to reducing fatal and serious injury crashes, contains information about common crash types across the states, and outlines treatments to address these crash types. *Road Safety Information Analysis: A Manual for Local Rural Road Owners (10)* also contains information on crash data collection and analysis techniques specifically applicable to local and tribal areas with limited resources.

Furthermore, programs such as LTAP and the Tribal Technical Assistance Program (TTAP), state DOT local offices, the State Highway Safety offices, and so forth, provide training, information, support, and guidelines for collecting and analyzing data.

Physical and Behavioral Aspects of Crashes

The severity of crashes (critical, serious, or property-damage-only), road conditions, lighting conditions, first harmful event, and so on, are some of the physical aspects of crashes that can be collated from crash reports. Several behavioral factors such as the driver's gender and age, alcohol-/drug-related crashes, safety device use, and speed can also be acquired from crash reports. The Governor's Highway Safety Association also provides information about behavioral safety issues. Physical and behavioral analyses of crash data are performed to identify ways to reduce fatal and incapacitating injury crashes.

In 2016, the Wind River Highway Safety Summit held at WRIR raised safety issues regarding driver behavior, presenting crash statistics and causative factors. The summit included an exclusive training session about alcohol and drug abuse, and treatments. In this session, how lessons learned from substance abuse prevention can be applied to traffic safety was discussed.

Application of the Crash Data Analysis

Crash analysis is performed to determine the common factors across a number of crashes. It also provides decision makers with critical information on what improvements are required at the crash locations.

Crash data must be acquired that represents an extended and recent period. Usually, 5–10 years of crash data provide enough information to identify high-risk crash locations. Accurate roadway locations along with roadway geometrics and pavement conditions should be included in the crash analysis. Geographical information systems (GIS) can be used to determine the precise locations of the crashes. After acquiring all of the necessary data, the analysis and subsequent ranking proceed using the crash data.

Each segment is ranked from the largest number of crashes per hotspot to the least number of crashes (*2*). Based on this ranking, the selected topmost high-crash routes are proposed for a Level I field evaluation.

Crash Rate Calculation. Depending on the availability of traffic volume data, the crash rate can be calculated to ensure a more accurate assessment of high-crash occurrence (*2*). Either a crash rate by vehicle miles traveled or a crash rate by route length can be obtained using the formula from FHWA.

Professional Examples

The methodology was successfully implemented on WRIR, SWO, SRST and YST. The findings of the implementation were summarized in comprehensive reports and submitted to all four tribes.

WRIR Crash Data Analysis. An SHSP incorporating the findings from the five-step methodology was developed for WRIR in 2012 (*11*, *12*). This safety plan has already been implemented and resulted in securing funding for several safety projects. Crash data from WYDOT and WRIR law enforcement were compiled through the collective efforts of tribal transportation, Wind River law enforcement, WYDOT and WYT^2/LTAP (*2*). After the implementation, there was reduction in crashes (see Figure 1). On average, the fatal crashes per year and the total crashes per year were reduced from 8 to 3 and 359 to 334, respectively.

Implementation of the five-step program resulted in implementing signs and pavement markings. The benefit–cost ratios for signs and pavement markings ranged from 8 to 20 (*12*).

The toolkit developed in this study provided guidelines to implement the methodology in a systematic way for the specific needs of tribes across the United States.

SWO Crash Data Analysis. After its successful implementation on WRIR, the methodology was executed on SWO roads. An initial ranking was performed based on GIS maps with the crashes overlaid on the roadways. The top 35 routes were subsequently selected for a Level I field evaluation.

Level I Field Evaluation

A Level I field evaluation is a systematic way to develop a ranking based on roadway conditions. Tribal leaders and any transportation supervisor within the community are eligible to perform the Level I field evaluation. This section is intended to assist users in forming the field review team, conducting the field evaluation, as well as providing guidance and resources to identify high-risk crash locations.

The first step of the five-step process entitled "compile data and crash data analysis" requires analysis to be carried out on all roads in the study area, and then 1-mi segments that have had three or more crashes in last 10 years are selected. The Level I field evaluation is performed on these selected roadway segments.

Level I Field Evaluation Review Team

A field review team should consist of at least one tribal member, one potential transportation expert such as member of LTAP, TTAP, or a transportation department, and one BIA engineering consultant to ensure that all aspects of the roadway segments can be adequately assessed.

Application of a Level I Field Evaluation

The roads are evaluated at 1-mi segments and given a score of 0–10: 0 being the most dangerous and 10 being the least dangerous (*2*). The descriptions of the five categories used for rating the road segments are as follows.

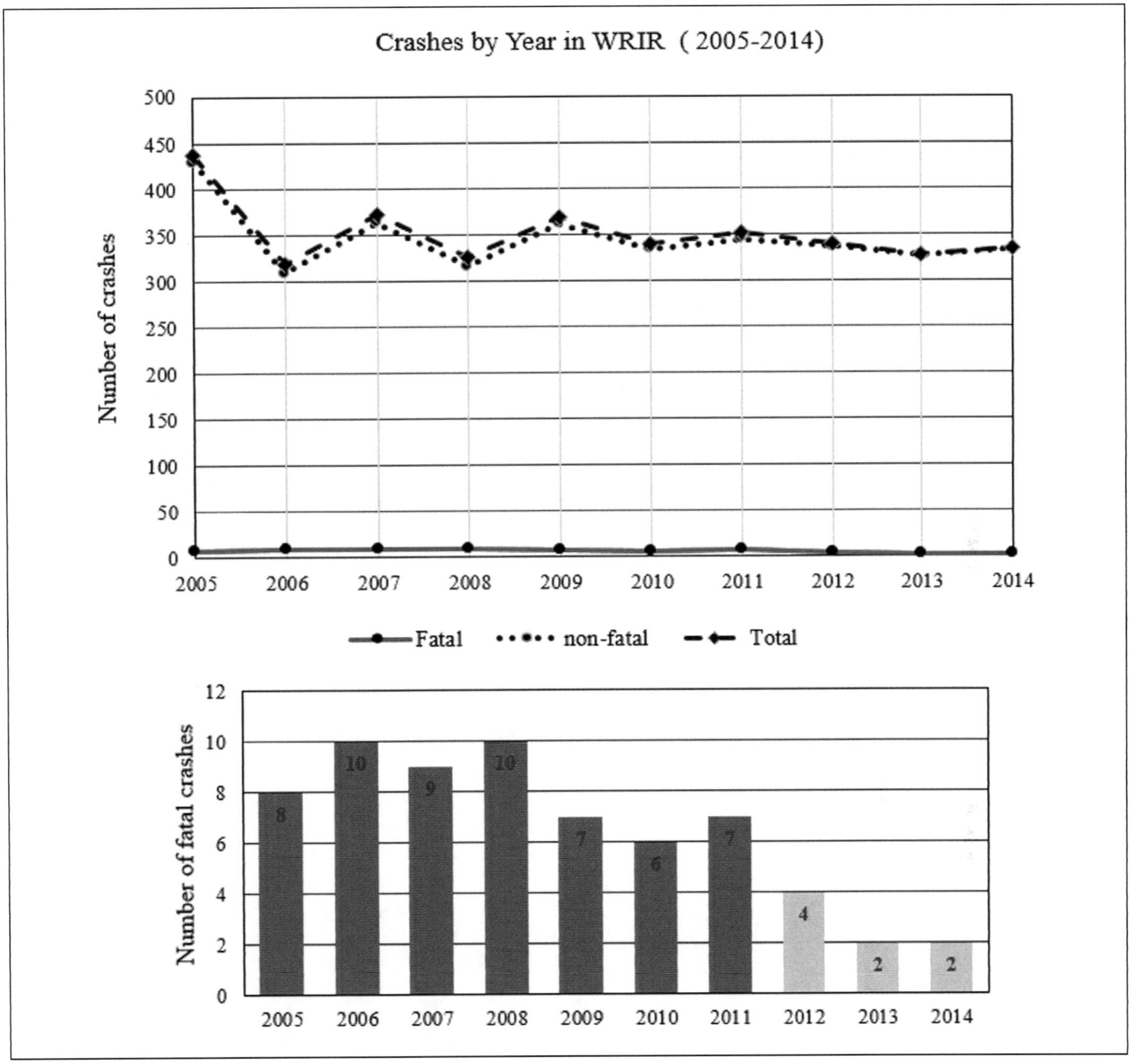

Figure 1. Total crash data and fatal crash data before and after SHSP in the WRIR.

1. *General:*
   - Presence of sharp horizontal or vertical curves.
   - Poor visibility that could result in safety problems.
   - Pavement defects that could result in safety problems.
   - Ponding or sheet flow areas that could result in safety problems.
   - Presence of loose aggregate/gravel that could cause safety problems.
2. *Intersection and railroad crossings:*
   - Intersections free of sight restrictions that could result in safety problems.
   - Intersections free of abrupt changes in grade or conditions.
   - Presence of advanced warning signs when intersection traffic control sight restrictions exist.
   - Presence of railroad crossing signs at railroad crossing approach.
   - Presence of railroad advanced warning signs when crossing sight restrictions exist.

- **FHWA Maintenance of Signs and Supports.** The FHWA prepared this guide in 2010. Sign principals, types of sign, installation of signs, signs location etc. are discussed in this guide to help local agency practitioners and maintenance staff in ensuring that their agency's signs are maintained to meet road user needs. Section 8 discusses inspection methods and offers maintenance staff a sign inspection checklist that is very helpful for conducting routine inspection.
  An electronic copy of the manual is available here:
  https://safety.fhwa.dot.gov/local_rural/training/fhwasa09025.

- **Vegetation Control for Safety.** This guide was made by FHWA in 2008. It aids local agency maintenance staff in identifying locations where vegetation control is essential to enhance traffic and pedestrian safety. The main goals of vegetation control include:
  - Keeping signs visible to drivers.
  - Keeping road users (vehicles, bicycles and pedestrians) visible to drivers.
  - Improving visibility of livestock and wildlife near the road.
  - Helping pedestrians and bicyclists see motor vehicles.
  - Keeping sidewalks and pedestrian paths clear and free from overhanging vegetation.
  - Removing trees close to the roadway which could result in a severe crash if hit.
  - Improving winter road maintenance in snow and ice areas.
  - Helping drainage systems function as designed.

Figure 2. Useful documents to evaluate roadway segments (FHWA).

- Vegetation and other obstructions restricting sight distance at railroad crossing.
- Roadway approach grade at railroad crossing level enough to prevent snagging.

3. *Signage and pavement markings:*
   - Signing present at needed locations to improve safety.
   - Presence of advanced warning signage at needed locations to improve safety.
   - Presence of unnecessary signage that may cause a safety problem.
   - Effective signage for existing conditions.
   - Presence of ineffective pavement markings for present conditions.
   - Presence of old or faded pavement markings affecting the safety of the roadway.
   - Presence of needed delineators.
   - Presence of improper or unsuitable delineators.
4. *Fixed objects and clear zone:*
   - Clear zones free of hazards, non-traversable side slopes without safety barriers.
   - Presence of narrow bridges or cattle guards.
   - Presence of culverts with inadequate extensions.
5. *Shoulder and right-of-way:*
   - Standard shoulder width.
   - Slope greater than 3:1.
   - Presence of hazards along shoulder.
   - High rollover potential.

Every segment receives a total score (maximum 50) summed from the score of each individual category. Each evaluated segment is provided a rank between the lowest and highest scores. The segment with the lowest rank is considered to have the highest risk (*2*).

Data from Existing Plans, Documents, or Other Agencies

Various safety policies and planning documents serve as guidelines in evaluating the roadways based on the five categories. For example, in order to assign signage and pavement marking category scores, FHWA's

Maintenance of Signs and Supports guide (*16*) and *Manual on Uniform Traffic Control Devices (17)* can serve as useful resources. The former manual describes sign principals, types of sign, installation of signs, sign location etc., while the latter describes the national standards for all traffic control devices including road markings, highway signs, and traffic signals. Again, in evaluating the fixed object/clear zone category, *Vegetation Control for Safety (18)* document can also be used as a guideline to assign scores. Moreover, for assigning scores for the intersection category, FHWA's *Intersection Safety: A Manual for Local Rural Road Owners (19)* can be utilized. Another manual, FHWA's *Roadway Departure Safety: A Manual for local Rural Road Owners* provides information on identifying locations with historical or potential rural roadway departure crashes, and countermeasures to address such incidents at these locations.

Field Examples

Field examples illustrate how to provide Level I scores for the road segments based on the five road condition categories. To begin with, all segments are given a rating of 5 out of 10 as an average. Scores are assigned to a segment based on the effectiveness/condition of the particular road condition features within the category. An example of this can be observed in Figure 3. This figure presents a roadway with serious pavement surface distress. This roadway might receive a score of 1 in the general category due to numerous potholes. It would also receive a score of 4 in the pavement marking category because of a missing edge line. However, a score of 8 or 9 could be given to the other categories. Consistency in scoring should be maintained throughout the evaluation process. Twelve field examples of different road conditions are described in the toolkit to assist the roadway evaluation process.

Professional Examples

WRIR Level I Evaluation. After the crash data analysis, a Level I field evaluation was conducted on WRIR. The evaluation team comprised one member from WYT^2/LTAP, a tribal transportation member, and a BIA engineering consultant. Each individual evaluated the roadways and assigned ranks based on the scores provided for each category.

SWO Level I Evaluation. In SWO, after the crash data analysis, the tribe selected 21 roads to be evaluated. The evaluating team was formed from four individuals: a SWO transportation safety officer, and representatives from WYT^2/LTAP, Northern Plains TTAP, and South Dakota LTAP. They performed the evaluation using the same procedure as the WRIR Level I evaluation.

Combined Ranking

This section describes the third step of the safety toolkit. In this step, crash ranking and Level I rankings are combined to develop a list of the priority sites based on the urgency of the countermeasures needed.

To determine the final rankings, the numerical values of the crash rank and the Level I rank are summed. The segments are then sorted by the combined rank value, in increasing order. The segments assigned the smallest numbers are considered the most hazardous. From these segments, the roads with the smallest combined ranking value are considered for a Level II field evaluation to determine countermeasures; 10–15 roads should be selected for the Level II evaluation.

The rankings along with the selected roads are submitted for review and approval of the tribes prior to proceeding with the Level II evaluation (*2*). The tribes have the option of including more sites or adjusting the rankings based on their insights.

Professional Examples

WRIR Combined Ranking. After the crash ranking and Level I field evaluation on WRIR, a combined ranking was performed. With a list of all the segments ranked by the highest number crashes and the lowest Level I score; the two rankings were combined by sorting each route and adding the respective ranks for the corresponding segment. After the ranks were tabulated, the segments were again sorted in ascending order to attain the combined ranking.

SWO Combined Ranking. The combined ranking on the SWO Reservation was performed after the crash ranking and Level I field evaluations were completed. The crash and Level I rankings were matched for every 1-mi segment for each route. The respective ranks for the corresponding segments were then added.

Once these were all totaled, the segments were sorted from the smallest to largest combined rank value. The road segments with the lowest score were used to select the roads that would be evaluated for safety improvements.

Level II Field Evaluation

The purpose of the Level II field evaluation is to identify causative factors on each road section and select corresponding countermeasures. This evaluation is performed on roadways that have been identified as high-risk locations based on the combined rankings.

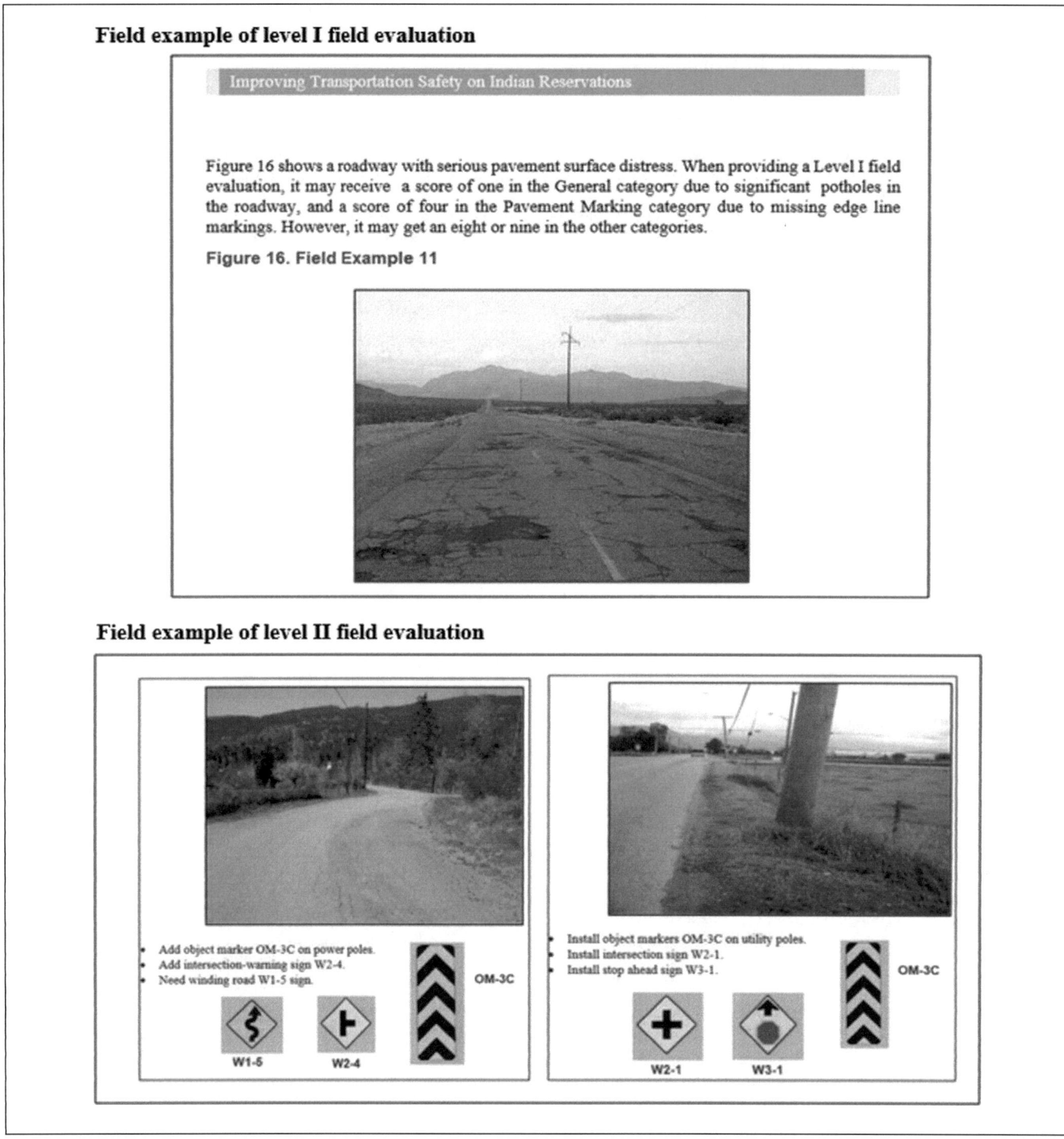

Figure 3. Field example of Level I and Level II field evaluation.

Once the tribes have identified their priority sites, an evaluation team performs the Level II field evaluation on each of the selected routes. Further information such as traffic speed, congestion levels, traffic counts, reviews of behavioral factors, and other casual factors that may influence the judgment of safety countermeasures may need to be collated (*2*).

Countermeasures should be considered based on the types of crashes and their frequency at the locations. Low-cost safety improvement countermeasures such as

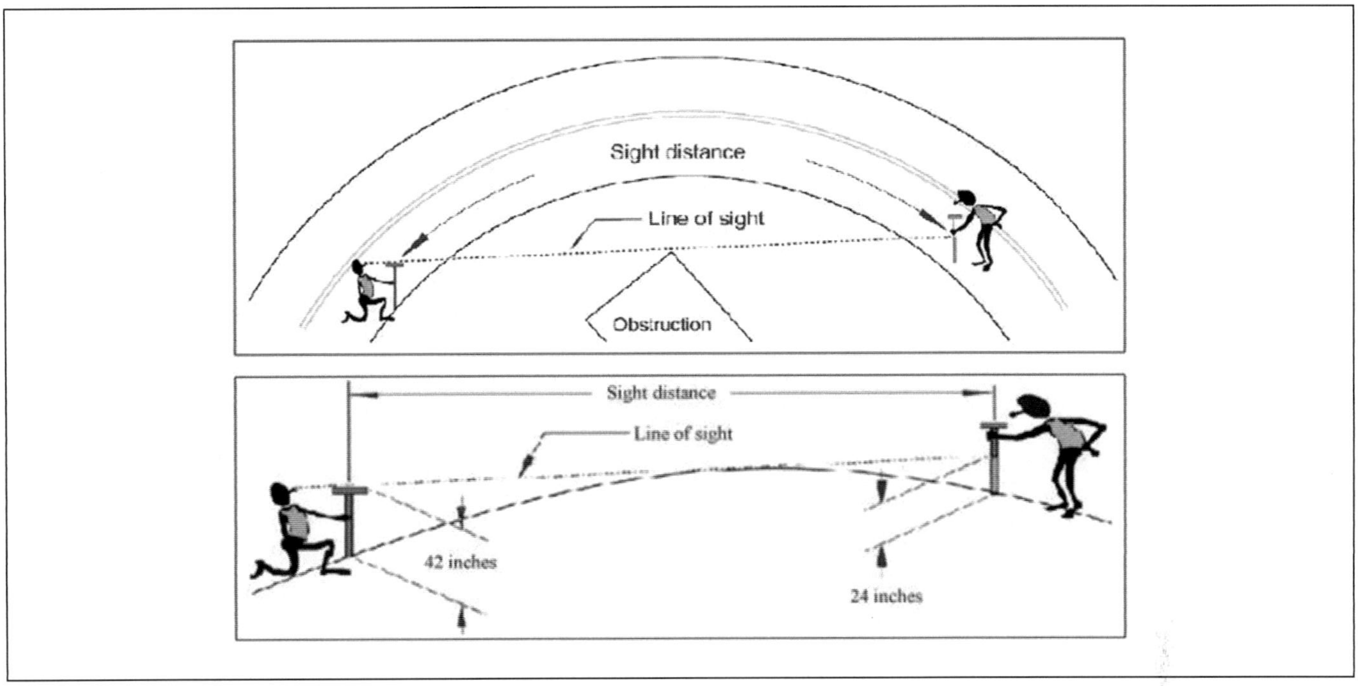

Figure 4. Measuring stopping sight distance for horizontal and vertical curves.

installation of advanced warning signs, chevrons at curves, delineators, and pavement markings can be used to reduce the number of crashes. A list of proven countermeasures and respective crash reduction factors can be obtained from FHWA safety websites (*13*).

In order to maximize time and resources, as well as to maintain consistency in evaluation, the Level I and Level II field evaluations should be performed simultaneously. This may seem perplexing in light of the purpose of Step 3 (combined ranking), however sometimes tribes want to improve all roadways involved in the safety evaluation, not just the segments or roadways that are highlighted in the combined ranking analysis.

Guidelines for Level II Field Evaluation

The toolkit provides helpful instruction during the Level II field evaluation for which horizontal curves, vertical curves, and other components may be a safety concern. The following sections can be used as guidelines to evaluate horizontal curves as well as measuring stopping sight distance.

Horizontal Curve Evaluation. The WYT^2/LTAP Center developed a simple procedure to measure a curve's radius in the field. In this procedure, a 100-ft rope is used that has a mark at 50 ft. After laying it on the shoulder of the road and pulling it tight, at the 50-ft mark – the distance from the rope to the shoulder of the road – is measured. This measurement, from AASHTO's *Guidelines for Geometric Designs of Very Low-Volume Local Roads (20)*, will give the middle ordinate of the curve to measure the radius and degree of curvature. Measured radius and degree of curvature can be compared with the minimum requirements set out of the county fund manual.

Horizontal and Vertical Curve Stopping Sight Distance. Stopping sight distance on all horizontal and vertical curves is measured along the travel path of the vehicle using a driver's-eye height of 42 in., looking at an object 24-in. high. Figure 4 shows the procedure for measuring stopping sight distance for horizontal and vertical curves.

Safety Issues to Look For

Major safety issues an evaluator may observe when performing a Level II field evaluation are described below.

A. *Roadside features*
   a) Are clear zones free of hazards and non-traversable side slopes without safety barriers?
   b) Are the clear zones free of nonconforming and/or dangerous obstructions that are not properly shielded?

B. *Road surface – pavement condition*
   a) Is the pavement free of defects that could result in safety problems (e.g., loss of steering control)?

   b) Are changes in surface type (e.g., pavement ends or begins) free of poor transitions?
   c) Is the pavement free of locations that appear to have inadequate skid resistance that could result in safety problems, particularly on curves, steep grades, and approaches to intersections?
   d) Is the pavement free of areas where ponding or sheet flow of water may occur resulting in safety problems?
   e) Is the pavement free of loose aggregate/gravel, which may cause safety problems?

C. *Road surface – pavement markings*
   a) Is the road free of locations with pavement marking safety deficiencies?
   b) Is the road free of pavement markings that are not effective for the conditions present?
   c) Is the road free of old pavement markings that affect the safety of the roadway?

D. *Road surface – unpaved roads*
   a) Is the road surface free of defects that could result in safety problems (e.g., loss of steering control)?
   b) Is the road surface free of areas where ponding or sheet flow of water may occur resulting in safety problems?
   c) Is the road surface free of loose gravel or fines that may cause safety problems (control, visibility, etc.)?
   d) Are changes in surface type (e.g., pavement ends or begins) free of drop-offs or poor transitions?

E. *Signing and delineation*
   a) Is the road free of locations where signing is needed to improve safety?
   b) Are existing regulatory, warning, and directory signs conspicuous?
   c) Is the road free of locations with improper signing which may cause safety problems?
   d) Is the road free of unnecessary signing which may cause safety problems?
   e) Are signs effective for existing conditions?
   f) Can signs be read at a safe distance?
   g) Is the road free of signing that impairs safe sight distances?

F. *Intersections and approaches*
   a) Are intersections free of sight restrictions that could result in safety problems?
   b) Are advance warning signs installed when intersection traffic control cannot be seen a safe distance ahead of the intersection?

G. *Special road users, railroad crossings, consistency*
   a) Are travel paths and crossing points for pedestrians and cyclists properly signed and/or marked?
   b) Are bus stops and mail boxes safely located with adequate clearance and visibility from the traffic lane?
   c) Is appropriate advance signing provided for bus stops and refuge areas?
   d) Are railroad crossing (cross bucks) signs used on each approach at railroad crossings?
   e) Are railroad advance warning signs used at railroad crossing approaches?
   f) Are railroad crossings free of vegetation and other obstructions that have the potential to restrict sight distance?
   g) Are roadway approach grades to railroad crossings flat enough to prevent vehicle snagging?

Field Examples

Figure 3 presents examples that may be similar to what would be observed and noted in an actual Level II field evaluation.

Professional Examples

WRIR Level II Field Evaluation. Generally, the state and U.S. highways within the reservation were in good condition and required few countermeasures. Many of the countermeasures included replacing or adding object markers on guardrails or other semi-hazardous objects. Chevrons were suggested on sharp horizontal curves where there was a high prevalence of run-off-the-road crashes. Widening of shoulders, and potentially adding rumble strips in locations with a high prevalence of rollover crashes were also requested. In a few locations, warning signs needed to be updated or added where there was a high presence of intersection-related crashes. Due to the high occurrence of wild animal crashes, more animal warnings signage was also recommended.

Data from Existing Plans, Documents, or Other Agencies. There are lots of safety policies and planning documents available that might be useful when performing a Level II field evaluation: These documents can be obtained from the FHWA's Office of Safety website and the FHWA Tribal Road Safety Audit Case Studies provide examples and guidelines for implementing road safety audits (RSAs) (*14*). Moreover, National Highway Institute offers a course on RSAs and RSA reviews.

Benefit–Cost Analysis

Based on the selected countermeasures, a benefit–cost analysis is performed to determine the most effective actions in reducing crashes at the minimal cost. This analysis considers both the costs and anticipated benefits of the countermeasures to determine their effectiveness. The following sections explain how to determine the components of a benefit–cost analysis as well as to calculate the benefit–cost ratio (BCR) in order to evaluate the economic effectiveness of the countermeasures.

Components of the Benefit Cost Analysis

Several factors such as anticipated benefit, crash cost, crash reduction factors, construction costs, environmental costs, planning and design costs, ongoing maintenance costs, service life and salvage value, interest rate and so forth, might affect decisions in selecting the countermeasures. Therefore, these components need to be estimated in a benefit–cost analysis. Standard crash costs for fatal, injury, and property-damage-only crashes provided by DOTs can be used to determine the crash costs of a particular high-risk area. FHWA and several DOTs also provide cost-reduction factors for different improvement programs, which can be used in the analysis.

Calculating the Benefit Cost Ratio

An incremental benefit–cost ratio method is used to compare the economic effectiveness among mutually exclusive countermeasures. In this method, incremental cash flow between the countermeasures, that is, the differences in benefits and costs between the countermeasures, is calculated, and then the ratio of the equivalent worth of incremental benefits to that of the incremental costs is determined.

Once the economic analysis is completed, a list of improvements with the BCR is provided to the tribes for their review and approval. After deciding which improvements they want to undertake, they can determine what resources they want to allocate to these projects.

Professional Examples

After the safety improvements were identified, the benefit–cost analysis was conducted on WRIR. The crash reduction factors (CRF) of the countermeasures were assigned according to the data provided by FHWA. Cost estimates were made based on WYDOT 2011 bid tabulations and other WYT^2/LTAP resources such as similar safety improvement projects. The total cost was calculated for all roadways and then compared to an overall benefit in crash reduction for the entire roadway. The benefit was calculated based on societal crash costs and represents the cost savings of a reduced number of crashes. A value was assigned to each type of crash severity which was obtained from WYDOT. The roads had a BCR ratio ranging from 2 to as high as 399. A closer look at the roads with a ratio over 100 reveals that many of the improvements are very low-cost but the benefit from the lives saved and injuries prevented is extremely significant.

Data from Existing Plans, Documents or Other Agencies

The *Highway Safety Manual* (Part B) and *Highway Safety Improvements Program Manual* provide details about the economic evaluation of safety countermeasures (*14*, *15*). In addition, the desktop reference for CRF-FHWA and the Crash Modification Factors Clearinghouse provide details to estimate the benefits based on the CRF and crash modification factors, respectively (*13*). Using these resources, the WYT^2/LTAP Center programed an Excel worksheet to calculate the BCRs for the safety countermeasures as shown in Figure 5. This figure provides screenshots of the programmed spreadsheets that include general site information, crash and countermeasure inputs, and costs and benefits for each countermeasure.

Summary and Conclusion

The safety toolkit provides a five-step methodology to enable the tribes to identify low-cost road safety improvements and allocate funding for these improvements. It also provides information about many useful resources in conducting the methodology and step-by-step examples of how to do so. This toolkit was implemented on multiple Indian reservations and showed great promise in reducing the high number of fatal crashes on the reservations roadways.

The paper has elaborately described each step of the toolkit, specifically, how the steps can be accomplished, field examples, professional examples implemented on WRIR and SWO, and resources for learning more about the steps. The first step in conducting a safety evaluation is compiling the available data and determining high-risk crash locations. After compiling and analyzing crash data from various sources, a ranking is established based on the high-crash locations. From this ranking, a list of roadways is proposed for field evaluation. Then each roadway is provided with a score by the field review team based on five categories: general category, intersections, signage and pavement marking, fixed objects and clear zones, shoulder and right-of-way. After that, each segment is provided with a rank from the lowest to the highest score. These two rankings are then combined to provide a list of proposed roadways for safety improvements. After the combined ranking, a Level II field evaluation is performed to identify causative factors on each road section and select corresponding countermeasures.

Input the general and site information into the table

General Information	
Analyst	Trenna T.
Agency/Company	WYTT
Project	FPIR
Date Performed	8/8/2016

Site Information	
Facility	
Road	BIA Route 1
Analysis Time Period	2005 to 2014
Analysis Year	2016
Segment Length (mi.)	78

Benefit to Cost Analysis Input Menu

	Number of Crashes			Countermeasures				
Road Segment	*Fatal*	*Injury*	*PDO*	*A*	*B*	*C*	*D*	*E*
BIA Route 1	5	32	7	2	3	5	8	17

Crash Cost Input Menu

	Countermeasures	Crash Type	Crash Reduction Factors			Cost	Service Life
			Fatal	Injury	PDO		
1	Install guide signs (general)	All	15%	15%	15%		5
2	Install advance warning signs (positive guidance)	All	40%	40%	40%	$9,000	5
3	Install chevron signs on horizontal curves	All	35%	35%	35%	$6,900	5
4	Install curve advance warning signs	All	30%	30%	30%		5
5	Install delineators (general)	All	11%	11%	11%	$234,000	4
6	Install delineators (on bridges)	All	40%	40%	40%		4
7	Install edgelines, centerlines and delineators	All	0%	45%	0%		4
8	Install centerline markings	All	33%	33%	33%	$5,400	2
9	Improve sight distance to intersection	All	56%	37%	0%		15
10	Flatten crest vertical curve	All	20%	20%	20%		15
11	Flatten horizontal curve	All	39%	39%	39%		15
12	Improve horizontal and vertical alignments	All	58%	58%	58%		15
13	Flatten side slopes	All	43%	43%	43%		15
14	Install guardrail (at bridge)	All	22%	22%	22%		10
15	Install guardrail (at embankment)	All	0%	42%	0%		10
16	Install guardrail (outside curves)	All	63%	63%	0%		10
17	Improve guardrail	All	9%	9%	9%	$18,000	10
18	Improve superevlevation	All	40%	40%	40%		15
19	Widen bridge	All	45%	45%	45%		15
20	Install shoulder	All	9%	9%	9%		5
21	Pave shoulder	All	15%	15%	15%		5
22	Install transverse rumble strips on approaches	All	35%	35%	35%		3
23	Improve pavement friction	All	13%	13%	13%		5
24	Install Rumble Strip	All	32%	32%	32%		10
25	Speed Study	All	10%	10%	10%		10

Calculate B/C Ratio

Calculation

	Countermeasures					
	A	*B*	*C*	*D*	*E*	Combined
Cost	$18,000.00	$13,800.00	$585,000.00	$27,000.00	$18,000.00	$661,800.00
Benefit	$5,784,800.00	$5,061,700.00	$1,590,820.00	$4,772,460.00	$1,301,580.00	$11,401,452.21
B/C Ratio	321.38	366.79	2.72	176.76	72.31	17.23

Figure 5. Excel worksheet to calculate the benefit–cost ratio.

Economic analysis is the fifth step in this toolkit and provides crucial information for the decision makers to prioritize projects and select appropriate safety countermeasures that can achieve the greatest economic effectiveness.

Successful implementation on WRIR and SWO Reservation reflects the adaptability of the five-step process to different Indian reservation areas. Strong cooperation and collaboration among the various stakeholders and tribal members accelerated the success of the program in WRIR and SWO. A high benefit–cost ratio of the countermeasures implemented in WRIR indicate that small improvements on these rural roads have a significant impact on the number of fatal and serious injury crashes proving the effectiveness of low-cost countermeasures.

The methodology developed in this study could be implemented on all reservations across the United States. Some minor modifications to this methodology may be required to address unique challenges in different reservations. The developed safety toolkit will be made available to all tribes by FHWA.

Acknowledgments

The WYT^2/LTAP center at the University of Wyoming provided extensive resources to assist in the development of the safety toolkit. The tribal leadership from WRIR and SWO tribes was very supportive and active in its implementation. Special acknowledgment goes to WYDOT and FHWA that provided the resources to make this research possible.

Author Contributions

The authors confirm contribution to the paper as follows: study conception and design: SN, TT, KK; data collection: SN, TT; analysis and interpretation of results: SN, TT, KK; draft manuscript preparation: SN, KK. All authors reviewed the results and approved the final version of the manuscript.

References

1. NHTSA. *Traffic Safety Facts 2006 Data.* https://crashstats.nhtsa.dot.gov/Api/Public/ViewPublication/810809. Accessed July 5, 2017.
2. Shinstine, D., and K. Ksaibati. Indian Reservation Safety Improvement Program. *Transportation Research Record: Journal of the Transportation Research Board*, 2013. 2364: 80–89.
3. EveryCRSReport. *Highways and Highway Safety on Indian Lands.* https://www.everycrsreport.com/reports/R44359.html#Content. Accessed July 5, 2017.
4. Centers for Disease Control and Prevention, National Center for Injury Prevention and Control. *Web-Based Injury Statistics Query and Reporting System (WISQARS)* [Online]. https://www.cdc.gov/injury/wisqars/index.html 2016.
5. *National Tribal Transportation Safety Summit Report.* http://flh.fhwa.Dot.Gov/programs/irr/safety/documents/summit-2009-11.Pdf. Accessed July 5, 2017.
6. *Chapter 12: Transportation Serving Federal and Tribal Lands - 2015 Conditions and Performance.* U.S. Department of Transportation, Federal Highway Administration, Washington, D.C.https://www.fhwa.dot.gov/policy/2015cpr/chap12.cfm. Accessed October 25, 2017.
7. *Tribal Safety Plans.* U.S. Department of Transportation, Federal Highway Administration, Washington, D.C. https://flh.fhwa.dot.gov/programs/ttp/safety/plans.htm. Accessed July 5, 2017.
8. Bailey, L., and D. Huft. Improving Crash Reporting: Study of Crash Reporting Practice on Nine Indian Reservations. *Transportation Research Record: Journal of the Transportation Research Board*, 2008. 2078: 72–79.
9. Ksaibati, K., B. Evans, and D. Shinstine. Implementation of Wyoming Rural Road Safety Program. *Transportation Research Record: Journal of the Transportation Research Board*, 2015. 2472: 109–116.
10. FHWA; Road Safety Information Analysis: A Manual for Local Rural Road Owners. https://safety.fhwa.dot.gov/local_rural/training/fhwasaxx1210/s1.cfm
11. Shinstine, D., K. Ksaibati, and F. Gross. Strategic Safety Management Plan for Wind River Indian Reservation. *Transportation Research Record: Journal of the Transportation Research Board*, 2015. 2472: 75–82.
12. Shinstine, D., K. Ksaibati, M. Carlson, and J. Smith. Roadway Safety on Indian Reservations. *TR News*, No. 294, September-October 2014, pp. 44–51.
13. *Tools and Technology – Safety.* U.S. Department of Transportation, Federal Highway Administration, Washington, D.C.https://safety.fhwa.dot.gov/tools/crf/resources/fhwasa08011. Accessed July 5, 2017.
14. *Improving Safety on Rural Local and Tribal Roads: Safety Toolkit.* U.S. Department of Transportation, Federal Highway Administration, Washington, D.C., 2017, pp. 1–5.
15. *Highway Safety Improvement Program Manual – Safety.* U.S. Department of Transportation, Federal Highway Administration, Washington, D.C.https://safety.fhwa.dot.gov/hsip/resources/fhwasa09029/. Accessed October 23, 2017.
16. Maintenance of Signs and Supports guide; US Department of transportation; FHWA-SA-09-025.
17. Manual on Uniform Traffic Control Devices; US Department of Transportation.
18. Vegetation Control for Safety; US Department of Transportation; FHWA-SA-07-018.
19. Intersection Safety: A Manual for Local Rural Road Owners; US Department of Transportation; FHWA-SA-11-08.
20. Roadway Departure Safety: A Manual for local Rural Road Owners; US Department of Transportation; FHWA-SA-11-09.
21. Geometric Designs of Very Low-Volume Local Roads; AASHTO, 2001.

The Standing Committee on Native American Transportation Issues (ABE80) peer-reviewed this paper (18-01482).

Sage Publications USA
sagepub.com

ISBN 978-1-5108-8177-8

9 781510 881778

CURRAN ASSOCIATES INC.
proceedings.com